DEFINING BIOLOGY
Lectures from the 1890s

DEFINING BIOLOGY

Lectures from the 1890s

Edited by Jane Maienschein

Harvard University Press
Cambridge, Massachusetts
London, England 1986

Library of Congress Cataloging in Publication Data

Defining biology.

Lectures given at the Marine Biological Laboratory in Woods Hole,
Mass., in the 1890s.
Includes index.
Contents: Introduction / Jane Maienschein—The mosaic theory of
development / Edmund Beecher Wilson—The hereditary mechanism
and the search for the unknown factors of evolution / Henry Fairfield
Osborn— [etc.].
1. Biology—Addresses, essays, lectures.
I. Maienschein, Jane.   II. Marine Biological Laboratory (Woods Hole,
Mass.).
QH311.D4 1986   574   86-307
ISBN 0-674-19615-5

CONTENTS

FOREWORD
Ernst Mayr

The meteoric rise of American biology is a fascinating historical phenomenon. Before the 1880s only two groups in the United States were interested in the living world. The interest of one of these, the
naturalists—nearly all of them located in museums and herbaria—was almost entirely restricted to the study of biological diversity. One of their
leaders, Louis Agassiz, was opposed to evolution up to his death in 1873.
The other group, the physiologists, usually teaching in departments of
medicine, likewise had very confined interests. The areas most characteristic of the biology of the twentieth century, like cytology, genetics, embryology, and neurobiology, as well as all of evolutionary biology, were
virtually ignored. Yet, thirty years later, the United States had produced a
group of world leaders in modern biology, including T. H. Morgan, E. B.
Wilson, E. G. Conklin, Frank Lillie, Ross Harrison, C. O. Whitman, H. S.
Jennings, and numerous others.

The research of the naturalists was dominated after 1859 by problems of
phylogeny. This branch of biology had a virtual monopoly in the United
States up to the end of the 1880s, represented particularly by Louis Agassiz's school, but also by the naturalists in the natural history museums in
New York, Philadelphia, and Washington. Morphology, as taught at some
colleges, prepared the way for a new era in American biology. Even
though William K. Brooks, who taught at Johns Hopkins from 1876 to
1908, was preeminently a morphologist with special interests in descriptive embryology, his students—among them Morgan, Wilson, Conklin,
and Harrison—became leaders of a new kind of biology. It was a biology
concerned with proximate causations, being experimental, analytical, and
explanatory. Professor Maienschein in her splendid introduction describes vividly how this new biology developed, and what important role
the Marine Biological Laboratory at Woods Hole played.

The role of marine biological stations in the origin of modern experimental biology cannot be exaggerated. None of them was as important as

the Naples station, where from the 1870s to the 1890s leading European biologists, particularly experimental biologists, met and stimulated each other and where the subsequent leaders of American biology became acquainted with the new trends in biology. In more recent years a special role was played by another biological station, the Biological Laboratory at Cold Spring Harbor, New York, where in the 1940s and early 1950s molecular biology came of age. As training centers for young biologists and as marketplace for the exchange of ideas, numerous other biological stations made important contributions, such as those of Plymouth, Roscoff, Banyuls, Kiel, Helgoland, Rovigno, and many others.

The opportunities provided by marine stations favor certain types of studies but also lead to a neglect of others. When Naples was founded in the early 1870s, it was in the heyday of interest in phylogeny and Ernst Haeckel's theory of recapitulation. No other problem seemed more urgent than to find the proper position on the phylogenetic tree for the numerous isolated invertebrate phyla. And, it was thought, the clues provided by ontogeny would give the answers. Hence, just about every worker at a marine station studied the embryology of marine invertebrates. And this tradition, transmitted by Brooks and his students, continued at first at Woods Hole.

Very soon, however, when the exciting experimental researches of Wilhelm Roux, Hans Driesch, and Oscar Hertwig became known, phylogenetic questions became less and less interesting. The focus shifted from evolutionary questions to those of proximate causation. This inevitably led to questions about heredity, an interest that was considerably sharpened by the controversies engendered by August Weismann's speculations. American biologists, who had spent a study period at Naples, brought this "new biology" from Europe to America.

From the very beginning, the Marine Biological Laboratory at Woods Hole served as a research station as well as an educational institution. The biologists, who for the rest of the year were more or less isolated at their colleges, should become acquainted with the state of the art in various subdisciplines of biology and learn about the current controversies at the frontier of biology. To achieve this objective, each summer a lecture series was provided. C. O. Whitman, who organized these lectures, was most anxious to present a fair balance of subjects, including each year at least some lectures on evolutionary biology, a rather minor interest of the experimental biologists. In the foreground of concern were embryological questions, particularly the organization of the egg cell in its earliest stages.

Is the development of the egg controlled by preformation or by epigenesis or by both? What is meant by preformation? Curiously, although the terminologies were changing and the study now took place at the cellular instead of the organismic level, the basic problems were still those that had preoccupied the embryologists in the eighteenth century. As we can now say with the benefit of hindsight, these questions could not be answered decisively until more was understood about genetics, until a cleaner separation was made between genotype and phenotype, and, in particular, until it was realized that an absolute difference exists between the instructions of the genotype (the DNA blueprint) and its translation into the proteins and other constituents of the phenotype. Yet, it was precisely the researches done by the Woods Hole biologists which clarified what questions ought to be asked, and this prepared the way for their answers.

Whitman insisted that the lectures were not to be devoted to detailed specific researches but rather to a survey of the leading issues of the day. It is quite fascinating to follow how the major emphasis of the lectures shifted from year to year, as is particularly evident when we compare the titles of the early years (1890–1893) with those of the last years (1897–1899). Some of the dominant themes of the early lectures had totally disappeared, replaced by entirely new themes—for instance, the nature of animal behavior.

Professor Maienschein presents a perceptive analysis of the interests of the Woods Hole biologists during this decade. She points out the increasing sophistication in the *Fragestellung* of the American authors, only few of whom (such as Jacques Loeb) were as radically one-sided as many of the Europeans. Even though solidly experimental, most of them, particularly those trained in the Agassiz school or by Brooks, had a background in natural history and at least some sympathy for the validity of evolutionary questions. However, Whitman's hopes for a unified science of biology that would pay equal attention to proximate and evolutionary questions did not materialize, and the schism in biology between the experimentalists and the naturalists became worse after 1900, with a synthesis delayed until the 1930s and 1940s.

The study of the MBL lectures is interesting and important for a number of reasons. First of all, they reflect an important stage in the history of American biology and reveal the reasons for some of the subsequent developments that would otherwise be rather puzzling. They show that some of the problems with which we are still struggling in the 1980s were already clearly seen by our forerunners ninety years ago. Indeed, some of their suggested solutions are amazingly up to date. But they also show that we have made real progress in the last one hundred years, and that some of the most controversial problems of the 1890s simply no longer exist, because decisive answers have since been discovered while certain alternatives were decisively refuted.

On first thought, one might imagine that lectures presented in the 1890s would be so dated that they are no longer of interest to a biologist of the 1980s. Reading these lectures, however, quickly dispels this notion. We must be grateful to Professor Maienschein for having selected some of the most important of these lectures and having provided us with a sensitive and informative introduction.

PREFACE
Jane Maienschein

The Marine Biological Laboratory in Woods Hole, Massachusetts, offered a series of evening lectures throughout the 1890s. As the first director, Charles Otis Whitman, wrote in his introduction to the published version of those *Biological Lectures,* they played a unique role for biology. By presenting critical issues of the day in general terms accessible to the whole community of researchers, the lectures could bring a much needed cooperative union among the various specialists. Whitman felt that such open interchange of ideas designed to transcend specialization would produce a vital community of American biologists. Indeed, Whitman was right.

These lectures can serve a similar purpose for biologists and historians today. This is, in part, because today's fields of embryology, genetics, biochemistry, cell biology, and behavior studies are the result of research programs pursued at the MBL in the 1890s. These research programs took the United States from a position as distant follower to the leadership role in biology in a very short time early in the twentieth century, and work at the MBL helped make this possible. The MBL provided more than the common root of diverse specializations. It also fostered an intense examination of how those specialized research programs affect and are affected by various broader questions of common interest. Precisely because the lectures addressed general questions and problems rather than presenting detailed research results, the lectures can teach us something about the background of today's specialties, about relations among those specialities, about how they fit into the overall concerns of biology.

The sense of community was important at the MBL and for the emergence of productive lines of research in American biology generally. Researchers gathered in Woods Hole each summer and found a group of people with related concerns, thus allowing them to move beyond the research isolation which most felt at their home institutions. The lecture series reflects the shared interests by addressing overlapping problems: of epigenesis and preformation, of the significance of past evolution, of heredity, of fertilization, of cleavage, of the importance of physiological processes or of environment for directing development. Moving from relatively descriptive cytological work to some manipulative experimental studies in the 1890s, the MBL community sought to understand what happens in development and how differentiation and organization arise and become established.

These *Biological Lectures* achieved a wide circulation through the 1890s within the small biological community, but they have remained relatively unavailable since. This volume of selected lectures should help to remedy that limited availability. The lectures that I have selected for reprinting here illustrate the vitality of the MBL community and center on papers by individuals who were active and thus responding to the shared interests there. Those individuals also served as leaders, by any standards, in biology in the twentieth century. These scientists have in each case recorded their findings and placed them in the context of broader problems of development. I could have chosen other papers or different themes, of course, but this set works to illustrate and demonstrate the importance of the core concerns.

The papers are not what scientists generally consider the "crucial" papers of biology, those which appear to have brought about the major changes in science. Rather, these are the papers which demonstrate the gropings of an important group of scientists at a critical time of change. They illustrate the processes of change and of defining new problems. Most of the problems have not been satisfactorily solved even yet, though biologists have certainly made some progress. Thus, this collection can serve to introduce modern students to some of the central problems of biology, and biologists to that period of scientific work when many of our current best assumptions about what biology should be like were being made.

I wish to thank the librarians at the MBL, especially Jane Fessenden and Ruth Davis, for their enthusiastic help in the preparation of this volume and for their exceptional generosity in providing open access to all materials in their collections. The archivists at the University of Chicago also provided useful materials. All archival passages from both libraries are quoted with permission. In addition, I appreciate the assistance offered at various stages by Joy Erickson, Richard Creath, and Ernst Mayr. National Science Foundation Grant #SES 85-10359 provided valuable financial support at a crucial time. Above all, it is really the spirit of the MBL from the 1890s which inspired this collection and which lingers still in a few laboratories and in a few individuals.

The photograph of Wilhelm Roux, on page 106, is reproduced courtesy of the Museum of Comparative Zoology, Harvard University, © President and Fellows of Harvard College. The photograph of Charles Otis Whitman, on page 218, is reproduced courtesy of the Joseph Regenstein Library, University of Chicago. All other photographs are reproduced courtesy of the MBL.

INTRODUCTION

INTRODUCTION
Jane Maienschein

The Marine Biological Laboratory opened in 1888, though a bit later than intended. One of the first students, Cornelia Clapp, who was to become a life-long supporter of the institution, arrived on time for the first session only to find carpenters still constructing the new building. No one had made living arrangements for the students; the director had not yet arrived, reportedly because of family illness; and the equipment, donated from the Annisquam Laboratory, remained side-tracked in a railway car somewhere along the way. In short, there really was no laboratory at the specified opening of that first summer session. Yet the students settled into boarding houses; the equipment made its way to Woods Hole; the director, Charles Otis Whitman (1842–1910), and the MBL opened officially on July 17, 1888. Aside from such persistent annoyances as stumbling at night over the many boulders in the paths (unimaginable in today's highly developed Woods Hole setting), that first session proceeded successfully. As Whitman later wrote, the MBL had begun with only seventeen "ids in its protoplasmic body—two instructors, eight students, and seven investigators (all beginners). The two investigators could be likened, with no great stretch of the imagination, to two polar corpuscles, signifying little more than that the germ was a fertile one, and prepared to begin its preordained course of development."[1]

That fertile germ had its origins in the Annisquam Laboratory, directed by Alpheus Hyatt (1838–1902) for the Boston Society of Natural History and the Woman's Education Association of Boston. The Annisquam Laboratory had its roots, in turn, in the Penikese Island school run by Louis Agassiz (1807–1873) and initially stimulated by Nathaniel Shaler

---

1. Whitman, Address to the MBL Corporation, 11 August 1903, Whitman Collection, MBL Archives.

(1841–1906). As Whitman pointed out repeatedy, the MBL was a lineal descendant of those two ancestors.[2]

The Penikese school was just that, a school. Intended to provide teachers with practical field experience in natural history, it began with financial backing from a wealthy New Yorker, John Anderson, who provided land and funds to build dormitories and laboratory space. Thus the Anderson School of Natural History opened its doors in 1873 to a collection of about fifty school teachers. Agassiz had so many applicants that he had to write to a few whom he had accepted in the early stages and ask them to withdraw in favor of better qualified candidates. The women students were very "schoolma'amy" and the "gentlemen are not a whit behind." according to one newspaper report. The opening proved quite a spectacular and newsworthy event as reporters and guests joined the students in New Bedford for the steamer trip to Penikese Island. Once there, they all celebrated the grand opening with a free picnic and an inspiring informal convocation. Only after the guests had departed for the mainland and the public attention had diminished did the students fully realize that they were on a virtually barren island about two-thirds of a mile long and one-third of a mile wide.[3] Fortunately, their regimen of work kept them sufficiently busy that they had no time to complain.

Some popular accounts give the impression that the students spent their days wandering about the island collecting things willy-nilly. It is true that the instruction was highly individualized, with each student spending a good part of each day exploring, collecting, observing, recording, and generally studying nature rather than books—as Agassiz

2. For discussion of this point see Jane Maienschein, "Agassiz, Hyatt, Whitman, and the Birth of the Marine Biological Laboratory," *Biological Bulletin* 168 Suppl. (1985): 26–34; Ralph Dexter, "From Penikese to the Marine Biological Laboratory at Woods Hole—The Role of Agassiz's Students," *Essex Institute Historical Collection* 110 (1974): 151–161; Dexter, "The Annisquam Sea-Side Laboratory of Alpheus Hyatt, Predecessor of the Marine Biological Laboratory at Woods Hole, 1880–1886," in Mary Sears and Daniel Merriam, eds., *Oceanography: The Past* (New York: Springer-Verlag, 1980), pp. 94–100.

3. For discussion of Penikese see, for example: Elizabeth Cary Agassiz, ed., *Louis Agassiz, His Life and Correspondence* (Boston: Houghton, Mifflin, and Co., 1885), chap. 25; Edwin Grant Conklin, "The Beginning of Biology at Woods Hole: Laboratory at Penikese Forerunner of M.B.L.," *Collecting Net* 2 (1927), no. 2: 1, 3, 6; no. 3: 7; Edward Sylvester Morse, "Agassiz and the School at Penikese," *Science* 58 (1923): 273–275; Albert Hagen Wright and Anna Allen Wright, "Agassiz's Address at the Opening of Agassiz's Academy," *The American Midland Naturalist* 43 (1950): 503–506; "Penikese Island," *Frank Leslie's Illustrated Newspaper*, 23 August 1873, pp. 377–378; Anonymous, *Penikese: A Reminiscence* (Albion, New York: Frank Lattin, 1895), p. 21; E. Ray Lankester, "An American Sea-Side Laboratory," *Nature*, 25 March 1880, pp. 497–499.

urged. But good books, not mere repetitive textbooks, also had their place. So did lectures. Agassiz invited a number of important biologists to address the group on a range of natural history topics. In fact, each day began with structured lectures, followed by an hour or so of dissection. Afternoons often brought freedom to roam and collect, but the evenings were spent attending further lectures, dissecting by candlelight, and then writing up notes from the day's work into the late night hours. Such a system obviously best suited those students capable of framing their own questions and following through with relevant collecting, but Agassiz and his invited speakers also helped to articulate appropriate problems.

The students attending that first year of an American marine school included Whitman and others who later spent time at the MBL, while Alpheus Hyatt reportedly also visited or lectured there.[4] The second summer promised equal success, with some students (including Whitman) returning for more advanced work. Cornelia Clapp (1849–1935) attended that second year, for example. Unfortunately, Louis Agassiz's death late in 1873 and his son Alexander's illness led to the closing of the school after that second year, as much because no one took the initiative to keep it going as for any other reason.

In 1877 Alexander Agassiz (1835–1910) opened his own private laboratory in Newport, Rhode Island, to which he invited an occasional visiting researcher until 1897, when deteriorating water quality forced its closing. Whitman visited there, for example, and began a coordinated project with Agassiz to study pelagic fishes. The younger Agassiz later complained, in refusing to support the MBL, that no one had joined him in providing such facilities for American researchers.[5]

Equally imporant, no one immediately took over the Penikese enterprise of teaching and providing practical experience in natural history. But by 1879 the Woman's Education Association and the Boston Society of Natural History decided that they needed a facility to instruct students, especially women, in this field. The Boston Society appointed Alpheus Hyatt as director and Balfour H. van Vleck (1851–1931), who had been a

---

4. Dexter, "From Penikese to the Marine Biological Laboratory," p. 161, provides a previously unpublished list of second-year students.

5. George Lincoln Goodale, "Alexander Agassiz (1835–1910)," National Academy of Sciences *Biographical Memoirs* 7 (1912): 291–334. Various documents in the Agassiz Collection at the Museum of Comparative Zoology Archives, Harvard University, such as letters from Agassiz to Hyatt of 30 May and 23 June, 1888, reveal Agassiz's reluctance to support the MBL and other such projects.

student at Penikese, as instructor. The summer school spent two years in Hyatt's house, then moved to a separate location nearby in 1881. There, the Annisquam Laboratory operated as a department of the Boston Society, with the continued support of the Woman's Education Association. This laboratory's purpose was instructional, in line with that of the Penikese school, whereas Alexander Agassiz's laboratory was more attuned to research. Hyatt's ideals helped to direct the lab, as did Van Vleck's particular notions of how to execute instruction. Yet the clear purpose of providing educational opportunities for science teachers and others came from the Boston Society of Natural History, for which Hyatt served as curator.[6]

At times the level of the students' commitment and preparation seemed depressingly low. As Mrs. Hyatt wrote to Alpheus when he was away on an expedition at sea, the group of students was very uninteresting, even tedious. They were essentially raw recruits, hopelessly elementary students, and they were beginning to drive poor Van Vleck to despair.[7] But the school also attracted such men as Thomas Hunt Morgan (1866–1945), who was to become an outstanding researcher and one of the backbones of the MBL.

In 1887 the Woman's Education Association decided that the project had succeeded and no longer need their support. They held the policy of seeding projects until they caught on, and then leaving them on their own.[8] The Annisquam project seemed a success. But Hyatt was tired and wished to develop an American marine laboratory on an independent basis, as an institution separate from the Boston Society of Natural History and from himself as director. He also felt that a new site would prove preferable to Annisquam, which was becoming polluted. Thus came the move to Woods Hole.

The reason for choosing Woods Hole lay largely with Spencer Fullerton Baird (1823–1887). For several years Baird had wanted his friend Hyatt to move his school to Woods Hole, which had purer water, more abundant marine life, a congenial setting, and, not coincidentally, the presence of the United States Fish Commission, which Baird headed. He wanted to build a marine research laboratory at the Fish Commission which would

---

6. Boston Society of Natural History's *Proceedings* provides discussion of goals and reports of activities.

7. Ralph Dexter, "Views of Alpheus Hyatt's Sea-Side Laboratory and Excerpts from his Expeditionary Correspondence," *The Biologist* 39 (1956–1957): 5–11.

8. Alpheus Hyatt, Boston Society of Natural History *Minutes*, 1887, pp. 3–4.

attract a community of researchers and students; Hyatt's school would prove a valuable complement to this project. Baird did attract cooperation from The Johns Hopkins University, which sent Professor William Keith Brooks (1848–1908) and some students to the Fish Commission, and from Princeton and Harvard. Yet Baird failed to gain the necessary financial support to attract other researchers and to establish a permanent biological research lab in the 1880s at Woods Hole.[9]

In 1887 Hyatt and the Boston Society found Woods Hole attractive indeed. Baird, who had befriended the Annisquam school by sending specimens, had urged a friend to buy land near the Fish Commission, which was held for the benefit of any educational institution that might build there. When the Trustees for the new Marine Biological Laboratory incorporated in 1888, they looked to Woods Hole as their site, and to the Fish Commission for further support.

Hyatt served as the first president of the MBL Trustees and encouraged the group to choose Brooks as the first director.[10] A professor of zoology at John Hopkins, Brooks was clearly one of the most visible of American zoologists. The Trustees hoped that the prestige of John Hopkins might also come with some financial backing from that school. In addition, Hyatt knew Brooks and felt that he might accept the position without salary. Brooks, who headed a small summer school for his own students called the Chesapeake Zoological Laboratory, had developed a working relationship with the Fish Commission and was thus familiar with Woods Hole. Each year John Hopkins had the right to send one or two students to do research there, in exchange for the university's one-time financial contribution to the Fish Commission. Brooks believed in research and wanted his advanced students to have practical laboratory experience. But he never gave any sign that he approved introductory instruction for other than advanced graduate students. The MBL's insistence on teaching as

9. Paul Galtsoff, *The Story of the Bureau of Commercial Fisheries Biological Laboratory, Woods Hole, Massachusetts.* (Washington, D.C.: United States Department of the Interior, 1962); Alpheus Hyatt, "Sketch of the Life and Services to Science of Prof. Spencer F. Baird," Boston Society of Natural History *Proceedings*, 1888, pp. 563–564.

10. MBL *Trustees Minutes* 1 (1888–1897): 11–13. On Brooks: Dennis McCullough, "W. K. Brooks's Role in the History of American Biology," *Journal of the History of Biology* 2 (1962): 411–438; Keith Benson, "William Keith Brooks (1848–1908): A Case Study in Morphology and the Development of American Biology," Ph.D. diss., Oregon State University, 1979; Benson, "American Morphology in the Late Nineteenth Century: The Biological Department at Johns Hopkins University," *Journal of the History of Biology* 18 (1985): 163–205.

well as research and its ancestors' emphasis on providing what Brooks would have regarded as essentially remedial field introductions did not appeal to him. Further, Brooks never became convinced of the wisdom of establishing a second lab in Woods Hole.[11] After some deliberations with MBL supporters, Brooks rejected the invitation to direct the MBL.

Immediately, on May 12, 1888, the Trustees offered the position to the only other American who had directed a biological laboratory—Charles Otis Whitman, then at the Allis Lake Laboratory in Milwaukee, Wisconsin. That unusual lab was founded by Edward Phelps Allis, Jr. (1851–1947), and essentially served as a teaching and research lab for him and a few other researchers. By May 18, Whitman accepted the offer to direct the MBL.[12]

The Woman's Education Association donated its equipment from Annisquam to the MBL and also helped the Trustees raise money for a new laboratory. With Van Vleck serving as first instructor, as he had at Annisquam, the MBL maintained connections with its founders. Yet Hyatt led the Trustees in making it clear that change was also in order, that the laboratory should offer both instruction and more advanced investigation, with instruction taking precedence if the Fish Commission succeeded in establishing itself as a research center.[13] Whitman should develop the lab as he saw appropriate. As Frank Rattray Lillie (1870–1947), Whitman's successor as director, later wrote, this decision worked out well, for in Whitman "the trustees had found a man not only fitted to carry out their purposes but possessing imagination adequate to transform their shadowy ideas, the zeal and determination required to give them form and substance, and the courage to face whatever difficulties might arise."[14]

During those first years, the Fish Commission proved very helpful in sharing specimens, providing seawater, a boat, nets, and so on. And the Fish Commission men (for unlike the MBL group, they were all men) visited and discussed projects. Clapp recorded that Whitman taught basic techniques and how to observe productively and to get results in morphological research. As she enthused about that first year, the year before the appearance of the soon-to-be-famous Wilson, Conklin, or Morgan, "The atmosphere of that laboratory was an inspiration, the days were

---

11. William Keith Brooks to President Gilman, no date, Gilman papers, Johns Hopkins University Manuscripts.

12. MBL *Trustees Minutes* 1 (1888): 23.

13. MBL *Trustees Minutes* reveal this arrangement in various places.

14. Frank Rattray Lillie, *The Woods Hole Marine Biological Laboratory* (Chicago: University of Chicago Press, 1944), p. 36.

peaceful and quiet; there were no lectures nor anything else to distract the attention from the work at hand."[15]

Although instruction meant introductory lab and practical work such as Agassiz's Penikese school had provided, it did not mean the sort of absolutely elementary work that the Annisquam lab had offered. Students were expected to have some preparation.[16] Thus, in the first years, work concentrated on the structure and life history of invertebrates, with considerable attention also to histological techniques, such as using microscopes, staining, fixing of specimens, and collecting materials to use in teaching. This last technique was important because the students mostly worked as school teachers.[17] Field work and careful observations were clearly emphasized, but so were the latest laboratory methods which could advance the observations. Those students with some training already who wished to pursue individual research projects were encouraged to do so. For the first year, lab work occupied virtually all the time for these independent investigators.

Cornelia Clapp, for instance, had enrolled as a student but was persuaded by Whitman that she had enough experience after her year at Penikese to undertake her own research under his guidance. Thus, she began work along lines Whitman considered important, on cleavage of toadfish eggs. Clapp became sufficiently enamored of research that she decided to take a Ph.D. in biology under Whitman at the University of Chicago, since her Ph.B. at Syracuse University, though satisfactory for her teaching position at Mount Holyoke, had not involved her in such advanced independent research work as she desired.[18]

Students at the MBL could later become investigators or even instructors, as many did in the 1890s and have continued to do. That particular blend of both instruction and investigation, originally endorsed because of the practical need to bring in money to run the laboratory and partly to avoid competition with the Fish Commission, became a life-long commit-

15. Cornelia Clapp, "Some Recollections of the First Summer at Woods Hole, 1888," *Collecting Net* 2 (1927), no. 4: 3, 10.

16. MBL *Trustees Minutes* (1888). Discussion in the Boston Society of Natural History's *Proceedings* 25 (1892): 282–283 reveals the changing climate in education which made the MBL requirement advisable. Teachers were no longer interested in elementary work and general courses by the 1890s, since they had already achieved a higher level of ability.

17. "Trustees' Report," *Annual Report*, 1888, pp. 19–20.

18. "Cornelia Maria Clapp," *Mount Holyoke Alumnae Quarterly* 19 (1935): 1–9. Her first publication from that MBL work appeared as "Some Points in the Development of the Toadfish (Batrachus tau)," *Journal of Morphology* 5 (1891): 494–502.

ment of Whitman's and one for which he had to wage numerous battles with those who would have given up the teaching. As he said in 1898 in his Presidential address to the Society of American Naturalists, instruction at first

> was accepted more as a necessity than as a feature desirable in itself. The older ideal of research alone was still held to be the highest, and by many investigators was regarded as the only legitimate function of a marine laboratory. Poverty compelled us to go beyond that ideal and carry two functions instead of one. The result is that some of us have developed an ideal of still wider scope, while others stand as they began by their first choice . . . On the basis of ten years' experience, and a previous intimate acquaintance with both types, I do not hesitate to say that I am fully converted to the type which links instruction with investigation.[19]

The Trustees' *Minutes, Annual Reports,* and letters reveal considerable debate about the proper role of instruction at the MBL, especially around 1902. But after 1902 it was agreed that some form of instruction would occur. The MBL had established its style and commitments during the 1890s under Whitman.[20]

CHARLES OTIS WHITMAN

The first director of the MBL remains one of the most underrated and understudied of the early American biologists.[21] Always quiet and reserved, to many he appeared far too serious. He was definitely never "one of the boys." Instead, Whitman stood as the stern but gentle and kind father figure for many of the young researchers at the MBL and at the University of Chicago, where he became head of the biology program in 1892. The epi-

19. Whitman, "Some of the Functions and Features of a Biological Station," *Science* 7 (1898): 37–44; MBL *Annual Report,* 1888, pp. 28–29, reveals some of the opposition to instruction; Carol Horgan, Archival Assistant at the MBL, has documented the changing attitudes toward instruction there.

20. W. D. Russell-Hunter, "An Evolutionary Century at Woods Hole: Instruction in Invertebrate Zoology," *Biological Bulletin* 168 Supp. (1985): 88–98, outlines one course of instruction.

21. Frank Rattray Lillie, "Charles Otis Whitman," *Journal of Morphology* 22 (1911): iv–lxxvii; Edward Sylvester Morse, "Charles Otis Whitman," National Academy of Sciences *Biographical Memoirs* 7 (1912): 269–288.

thet "sober and pious Yankee" seems particularly appropriate for this shy and sometimes unhappy man.

The son of farmers, Whitman grew up in Woodstock, Maine. He very soon developed a taste for bird collecting and for roaming through the woodlands near his home, generally preferring his own company to that of others. His uncompromising commitment to principle and the very early whitening of his hair often set him apart as an authority rather than a compatriot for his peers, although he was reportedly a loyal friend. His personality annoyed many throughout his life, but his stubborn refusal to compromise also doubtlessly made it possible for the MBL to develop into the successful enduring institution it became.

After high school, Whitman attended Bowdoin College, receiving his B.A. in 1868 with a largely classical education. Thereafter he served as teacher and principal at Westford Academy in Massachusetts from 1868 to 1872. He also substituted at English High School in Boston during the 1871–1872 school year and received a regular appointment from 1872 to 1875. There Whitman was attracted by Louis Agassiz and signed up for a summer at Penikese in 1873. He had continued his youthful ornithological interests with bird collecting and preparations but had never formally studied or taught natural history or biology. At Penikese, Whitman met Edward Sylvester Morse (1838–1925), who lectured to students on the natural history and embryology of molluscs.[22] Morse, who later played an influential role in obtaining a job for Whitman, was impressed by Whitman's careful and beautiful drawings, a skill for which Whitman received considerable acclaim and which he always cultivated in his own students. Whitman returned for a second summer at Penikese as one of the advanced students.

As Whitman later reported to his student at the University of Chicago, Wallace Craig (1876–1954), he first began scientific work in zoology under Louis Agassiz, but "did not really get under way until he worked with Leuckart on Clepsine in Germany."[23] In fact, the summers at Penikese helped convince him to pursue natural history studies. In particular, the zoology laboratory of Rudolf Leuckart (1822–1898) at the University of Leipzig attracted a number of American students. There Whitman re-

22. "Professor Agassiz's School of Natural History," *Popular Science Monthly*, 1873, 123–124.
23. Wallace Craig, memo to Frank Lillie about Whitman, 29 August 1910, University of Chicago Archives.

ceived his Ph.D. in 1878 for his meticulous study of the early developmental stages of the leech *Clepsine,* a parasitic organism in which Leuckart himself was much interested.[24] His earliest paper addresses suggestions that "precocious segregation" may in fact occur, as Edwin Ray Lankester (1847-1929) and Wilhelm His (1831-1904) had suggested, so that the egg cell already experiences a heterogeneous organization which undergoes "histological sundering" in the course of individual development. That work and subsequent investigation inaugurated the important tradition of cell lineage study at the MBL in the 1890s. As biographer Morse suggested, Whitman reminded "one of a German type of mind" with the meticulousness of that work.[25]

Upon return to the United States, Whitman spent one further year at English High School, then resigned. He applied to The Johns Hopkins University for a special position for graduates as a Bruce Fellow there and received an appointment for 1879-1880, with strong recommendations.[26] Before assuming the fellowship position, however, he left to accept the chairmanship of the biology department at the Imperial University of Toyko, for which Morse had recommended him. With that move he committed himself to a career in scientific research work and teaching in biology.

The Imperial University of Tokyo sought to establish a modern biological department and imported Americans, namely Morse and then Whitman, to direct the program after Thomas Henry Huxley (1825-1895) declined the invitation.[27] In Tokyo, Whitman had four students who completed the program, all four of whom became professors of zoology. Their fond reminiscences upon Whitman's death attest to his influence in Japan. He taught the students the latest histological techniques imported from Germany, he instructed them to draw carefully in order to produce reliable records of their observations, and he introduced them to what it meant to undertake scientific research. One student's later recollections

24. See Klaus Wunderlich, *Rudolf Leuckart* (Jena: Gustav Fischer Verlag, 1978), pp. 41-49, for a list of Leuckart's students; Charles Otis Whitman, "The Embryology of Clepsine," *Quarterly Journal of Microscopical Science* 18 (1878): 215-315.

25. Morse, "Whitman," p. 278.

26. Lillie, "Whitman," p. xix; The Bruce Fellowship, named after Johns Hopkins graduate student Adam Bruce, was established for post graduates to continue their studies in biology (Johns Hopkins University Manuscripts).

27. Lillie, "Whitman," pp. xix-xxiv.

of those days reveals Whitman's influence as well as his views of what good biological work required:

> We had under Professor Morse, only two courses of lecture, general zoology and evolution; and one laboratory work of comparative anatomy. After Whitman became our professor of zoology, Zeiss microscopes and microtomes were newly brought in. A new course in embryology was organized. Each student using his own microscope made experiment of embryology. As we were all rooming in a dormitory, we used to work in the laboratory till twelve at night. Professor Whitman was very industrious in his work, all day long he studied material under microscope. Professor Whitman's office was next to our laboratory and his working table in the office was situated as such that he could see us all, by pushing a door between his office and our laboratory. The door was always kept widely open and as consequence we felt somewhat restrained. At times, he was away, we began to talk and turned the scene quite noisy; but no sooner we have heard his foot step drew near than the noise was gone, stillness reigned again and all students seemed busy peeping microscopes unconsciously . . .
>
> Professor Whitman put much emphasis upon the microscopical study and did not seem to care mere collection of material at random.[28]

Whitman shared his equipment, books, and journals, even helping the students to translate from German and French into English. In Japan, he continued his own research on leeches, examining development and life histories of several species for comparison with his German *Clepsine.*

During the two years Whitman spent in Tokyo, he gained a reputation as the father of Japanese zoology.[29] Yet he left after several skirmishes with university authorities, most notably over publication of his students' papers. By the end of two years, each of his four students had produced a research paper. These Whitman submitted to the university science journal. When informed that only papers by professors could be published

28. Tomotaro Iwakawa, "Professor Charles O. Whitman," trans. Shigro Yamanouchi from the Japanese *Magazine of Zoology* 23 (1911): pp. 2–3, Whitman Collection, University of Chicago Archives.
29. Lillie, "Whitman," p. xx; Whitman, "Zoology in the University of Tokyo," unpublished manuscript.

there and that he should add his name as author to each of the student papers, Whitman's commitment to principle prevailed. He withdrew the papers and submitted them elsewhere. Three appeared in the *Quarterly Journal of Microscopical Science* and the fourth soon appeared in the university journal after all. When his contract expired in August 1881, Whitman declined an invitation to stay and left to return to the United States by way of Europe. As one of his students rather awkwardly reported, despite his conflicts with the university administration, "Professor Whitman loved Japan and sympathized with the Japanese. That his love and sympathy poured forth toward the Japanese in a degree far surpassed than any ever shown by any to us, were marvelously evidenced in the time of Russo-Japanese war."[30]

Before returning to the United States, Whitman stopped at the Naples Zoological Station from November 1881 to May 1882 and worked there as a personal guest of director Anton Dohrn (1840–1909). Officially, a researcher was supposed to be sponsored by an institution and to work at a subscribed table. The United States had taken no subscription when Whitman arrived as the first American at Naples, however, so Dohrn welcomed him as a guest. That Dohrn did not intend to allow this to become standard practice is evident from his insistence that the second American at Naples, Edmund Beecher Wilson (1856–1939), obtain permission to work at some subscribed table or other.[31] At Naples, Whitman studied the dicyemids (parasites of cephalopods), and, following the best standards of the German morphological tradition, he traced the development, life history, behavior, and classification of that form.[32] His careful work well repaid Dohrn's hospitality.

After Naples, Whitman had no job and remained uncertain about whether to pursue the Bruce Fellowship offered two years before at Johns Hopkins. He visited Leipzig for several months, then returned to the United States. There he received an appointment in 1882 as Assistant in Zoology at the Museum of Comparative Zoology at Harvard University, where he remained until 1886.

30. Lillie, "Whitman," p. xxiii for discussion; Chiyomatsu Ishikawa, "Professor Charles O. Whitman," in Yamanouchi, "Whitman," 1911, p. 16, and Katashi Takahashi, "My Old Professor Dr. Charles O. Whitman," 1911, in Yamanouchi, p. 23.

31. Edmund Beecher Wilson to President Gilman, 9 March 1883, student file, The Johns Hopkins University Archives.

32. Charles Otis Whitman, "A Contribution to the Embryology, Life-history, and Classification of the Dicyemids," *Mittheilungen aus der Zoologischen Stazion zu Neapel* 4 (1883): 1–89.

From 1886 to 1889 Whitman directed the Lake Laboratory in Milwaukee to instruct Edward Phelps Allis, Jr., and to conduct biological research. Allis, an independent man interested in biology, had been advised by the British physiologist Michael Foster (1836–1907) to pursue research rather than reading or visiting other people's laboratories. After being tutored for a year by a Johns Hopkins graduate, Henry van Peters Wilson (1863–1939), he sought a different arrangement. Whitman received the highest recommendation, and so Allis invited him to direct a research laboratory where Allis would take part in and learn from the work. Over the Lake Laboratory's eight-year existence, several researchers spent time there and produced published studies, especially in embryology.[33]

His experience at the Lake Laboratory undoubtedly influenced Whitman's life-long hope of establishing an inland "biological farm," as he called it, to complement the marine work of the MBL. His continued failure to convince donors and administrators of the value of such a farm proved a recurring frustration. He wrote in 1895 to Helen Culver, who donated one million dollars for biological work, that he expected her money to fund three projects: biological laboratories at the university, an inland marine station, and a marine biological observatory. The first and third were already organized, he pointed out, and said of the second that "the Experimental Station comtemplated is something wholly new and unlike anything thus far provided for in America or Europe."[34] Somehow the project never came to fruition, and the money all ended up at the University of Chicago rather than at the MBL or an experimental farm.

In addition to the Lake Laboratory itself, Allis also agreed to support the publication of an American journal for zoological work. This journal must remain independent of any particular society and from European direction, Whitman had urged. The *Journal of Morphology* was the result. As Whitman noted in the introduction to that journal, first published in 1897, "The mixed character and scattered sources of our publications are twin evils that have become intolerable both at home and abroad. The establishment of the Journal of Morphology may not be the death blow to these evils; but there is hope that it will, at least, relieve the more embar-

33. Lillie, "Whitman," pp. xxv–xxvii; Ernst J. Dornfeld, "The Allis Lake Laboratory," *Marquette Medical Review* 21 (1956): 115–166, esp. pp. 118–120.

34. Whitman to Helen Culver, 20 December 1885, p. 1, Whitman file, Lillie Collection, MBL Archives. Materials in the Presidential Papers, University of Chicago Archives, contain similar discussions and yet do not clarify what caused the changes of plans for Miss Culver's money.

rassing difficulties of the present situation."[35] Though the journal produced a financial loss for its first years, Whitman managed to persuade the publishers to persist until 1903. After a few years' lapse, the Wistar Institute took up publication again in 1908. In 1898 Whitman also began the *Zoological Bulletin* with William Morton Wheeler (1865–1937), who had worked at the Lake Laboratory and studied under Whitman at Clark University in Worcester, Massachusetts, where Whitman served as chairman of zoology from 1889 to 1892. Intended for shorter articles and reports, the *Zoological Bulletin* was seen as a complement to the *Journal of Morphology*. In 1890, the name changed to the *Biological Bulletin* in accordance with the desire to recognize a general discipline of biological research which combined zoology and botany. The editing and publication moved to the MBL.[36]

In 1889 Whitman had hoped to receive an appointment at Columbia University, but did not.[37] Instead, he went to the newly opened Clark University, whose graduate research orientation appealed to Whitman. He quickly began to attract his own students, who went with him to the MBL for summer work. Unfortunately, Whitman was one of many Clark faculty members who soon experienced displeasure with the administration over various issues. In 1892 he and others left Clark to become the distinguished core of the new University of Chicago biology program. The brief Clark stay was nonetheless important for Whitman. During those three years he was simultaneously interpreting for the department there, for the MBL, and for his journals and lectures series what biological work should be like, what problems were important, what methods were appropriate, and which investigators should be encouraged. He had tremendous influence in those years over the shape of American biology, with little competition except from Johns Hopkins. And the best students from Hopkins all found their way to Woods Hole and fell under Whitman's influence to some degree as well. His net of connections was spread wide, and loyal students such as Frank Lillie followed him from Clark to Chicago and to the MBL and back. Investigators at the MBL returned year after year and assumed leadership roles there at Whitman's request. MBL researchers published in the *Journal of Morphology*, the *Biological Bulletin*, and the

35. Lillie, "Whitman," p. xxvi.
36. The prospectus for each journal reveals the differences in goals.
37. Whitman letters to Alexander Agassiz, Agassiz Collection, Museum of Comparative Zoology Archives, Harvard University.

Biological Lectures series. Biology remained a small field prior to 1900, and Whitman exerted a powerful influence on most of the American researchers.

WHITMAN'S IDEALS FOR BIOLOGY

It was at the MBL that Whitman most directly revealed his ideals for American biology and most successfully played them out. In his *Biological Lectures* series, his *Annual Report to the Corporation,* and his letters to friends and colleagues he articulated his hopes and frustrations. "Specialization," "cooperation," and "independence" served as Whitman's watchwords for biology.

In his inaugural lectures for the *Biological Lectures* series in 1890, Whitman insisted that specialization was desirable, indeed necessary for biology. No one investigator could manage all of biology any longer. Specialization and organization had become "companion principles of all progress" and "the most important need of American biology," according to Whitman. American biology lagged far behind German work, he had lamented in 1889, partly because of the failure to specialize and cooperate.[38] People fear specialization, he acknowledged, but proper understanding could bring acceptance. Essentially, specialization in science was an expression of the principle of division of labor which societies of individuals all experience. Individual cells, he said, begin as independent units, each nearly like the others. Progressive development brings specialization, with mutual dependence, and resulting social organization. Division of labor brings union of the laborers, with the individual parts continually responding to their places in the whole. There is no pre-existing structure which determines what each individual cell (or person) shall become, but hereditary potentials influence the particular division of labor that occurs within the system. With higher degrees of specialization and organization comes higher rank in the "scale of life and intelligence," since organization depends on each part's sense of place in the whole, and knowledge of the whole presupposes knowledge of the parts as well. Some fear division of labor because they fear that the body—or science or whatever—will disintegrate into unrelated parts, but "there are centripetal forces that keep pace with the centrifugal ones; and the danger of any

38. Also discussed in Whitman, "Biological Instruction in Universities," *American Naturalist* 21 (1887): 507–519.

science flying into disconnected atoms is about as dreamy and remote as the dissolution of the earth itself."[39]

Specialization tends to run ahead of organization, so researchers must work cooperatively to keep organization in sight. A permanent national marine biological station, presumably the MBL with a solid financial basis and truly national support, would enhance this essential cooperation and organization by bringing together various specialists. Such a station was "*the* greatest desideratum of American biology."[40] The ensuing efforts to secure a solidly founded national station while ensuring independence from any one agency, university, or individual earned Whitman considerable heartaches and ultimately led him to withdraw from MBL administration. The move toward specialization without cooperation, and the insistence on analytical study of parts and not the whole in biology, similarly discouraged the idealistic and optimistic Whitman. Yet he persisted throughout the 1890s in pushing his ideas into effect.

In his second lecture to the MBL, also in 1890, Whitman elaborated what he saw as current specialty areas of research. Some seek to establish geneological relationships and a system of classification; others examine those forms that exist now. Comparative anatomy, paleontology, and embryology all work in parallel to find homologies, for example. Similarly, geological succession and embryological development, zoological gradations, and geographical distribution of animals also are parallel phenomena, each revealing the community of descent. Study of each runs parallel to study of others. Each informs the others. For any topic of research, one would have begun "with a special problem and found it to be the centre of inquiries, leading in all directions into the unknown. So it is with all special subjects in biology. The farther we pursue them the broader and more interesting they become. Nothing could be farther from the truth than the idea that such questions are isolated, and devoid of interest to all except the specialist."[41]

Even morphology and physiology, thought by many to represent two distinct and even divergent research traditions, were two aspects of the same thing, in Whitman's view. While one studies form, the other studies function of the same organism. Physiologists have ignored important problems such as the fundamental processes of heredity, variation, and

39. Whitman, "Specialization and Organization," 1890, p. 22.
40. Ibid., p. 24.
41. Whitman, "The Naturalist's Occupation," 1890, p. 52.

adaptation, and morphologists have only just begun to address some of the physiologists' traditional questions. Physiologists must learn to appeal to paleontology and especially to embryology as sources of evidence, because "the embryological series, often including free larval stages, furnishes one of the grandest fields for experimental study. Here the physiologist has an opportunity not only to study by experiment but also by direct observation and inference, and thus to join hands with the morphologist both in methods and results."[42] Thus, at the MBL Whitman developed a department of physiology in 1894, headed by Jacques Loeb (1859–1924), whom he imported from the University of Chicago. From the beginning the MBL included botany as well as zoology, though the latter clearly maintained its primacy. He also introduced new courses as they seemed of sufficiently general interest and substance.

Despite Whitman's stature as a leader of American biology, battles at the MBL over money and control of the laboratory climaxed in 1897 and 1902 and led to his discouragement and eventual withdrawal from his influential post there.[43] Though Whitman remained the official head of the MBL until 1908, he discontinued publication of the *Biological Lectures* in 1900 and of the *Journal of Morphology* in 1903, and in effect withdrew from the MBL and the University of Chicago after 1902, relinquishing control of the MBL to his assistant, Lillie, who became the official director in 1908.

Whitman had by that time returned to his beloved birds, concentrating on the evolution, behavior, and development of pigeons. His unhappy family life, including his apparent estrangement from his wife, Emily Nunn, and his difficulties with a pathologically shy and troubled son, led him increasingly to communion with his pigeons.[44] Indeed, his final paper to the MBL in 1898 on animal behavior is probably Whitman's masterpiece. As his biographer, Morse, said of that lecture, which he evidently heard,

42. Whitman, "General Physiology and Its Relations to Morphology," *American Naturalist* 27 (1893): 802–807. From the "5th Annual Report of the Director," MBL (1892).

43. Whitman to confidant Edwin Grant Conklin, a series of letters, Whitman Collection, MBL Archives.

44. Whitman's friends systematically avoided discussion of his family problems, as revealed in Lillie Collection, MBL Archives, and Whitman Collection, University of Chicago Archives. Pieces of letters hint that major problems existed, for example, Ishikawa, "Whitman," 1911, p. 18, reports that Mrs. Whitman and their son were not living with Whitman when Ishikawa visited in 1908. Letters from Mrs. William Keith Frost to Lillie, Whitman Collection, University of Chicago Archives, provide evidence of Whitman's frustrations.

Dr. Whitman gave one of his most interesting and delightful essays. The table of contents even is as enjoyable as the menu of a rich feast. The lecture is crowded with facts which reveal his wonderful powers of observation. General considerations regarding the origin of instinct, which he shows precedes intelligence, and weak points in the habit theory, etc., indicate his thorough knowledge of the various discussions which have been published. In short, a fair presentation of this luminous lecture would be impossible in this brief memoir. It may stand as a model for discourses of this nature.[45]

That work is supplemented by the posthumous publication of three impressive volumes on pigeons, painstakingly and loyally undertaken by Oscar Riddle (1877–1968), Whitman's student at Chicago who carried on Whitman's pigeon work, indeed carried on with Whitman's own pigeons after Whitman's unexpected death in 1910. Yet Whitman and his studies of pigeon evolution had come to be regarded as old-fashioned at the University of Chicago by this time. When Riddle took up the work, he was effectively instructed to leave Chicago.[46]

Whitman had been caring for his avian friends in their coops in his backyard during a sudden cold spell in Chicago in November 1910. He developed pneumonia and died shortly. Though his helpful assistant, Lillie, officially took over the Chicago department as well as the MBL, it became clear that Lillie did not share all of Whitman's commitments. He did nothing to help Riddle, for example, though Whitman had promised Riddle an official faculty position upon his return from a research visit to the Naples Zoological Station. Riddle received word that Lillie would not honor Whitman's commitment, partly because of departmental sentiments, and it was only through the intervention of Whitman's friend Albert Prescott Mathews (1871–1957), a physiologist, that Riddle obtained sufficient funds to remain in Chicago and to keep the pigeon colony alive for another year. The next year, Riddle and Whitman's pigeons departed Chicago for the Carnegie Laboratory at Cold Spring Harbor, a symptom of

---

45. Morse, "Whitman," p. 278.

46. Oscar Riddle, ed., *Orthogenetic Evolution in Pigeons* (vol. 1), and *Inheritance, Fertility, and the Dominance of Sex and Color in Hybrids of Wild Species of Pigeons* (vol. 2). Harvey A. Carr, ed., *The Behavior of Pigeons* (vol. 3) (Washington, D.C.: Carnegie Institution, 1919). Takahashi, "Whitman," 1911, pp. 24–25, reports that Whitman asked him repeatedly to stay and work on the pigeons since Whitman needed assistance. Whitman reportedly was spending most of his salary ($7,000) and most of his time on the pigeons.

the erosion of support at Chicago for Whitman's program of behavioral research and for his ideas about biology.

LEADING PROBLEMS: PREFORMATION AND EPIGENESIS

Despite the loss of support in the 1900s, Whitman's ideals played a major influential role in shaping American biology in the 1890s. The very idea of having a series of biological lectures to address central issues of the day, as well as the particular speakers and subjects chosen, reveal Whitman's personal stamp.

The focus of interest of the *Biological Lectures* shifted from year to year as new discoveries brought new questions, but some themes underpinned discussion throughout the 1890s. Most notably, questions about the significance of heredity and evolution for development, and related questions about the significance of cell cleavage for differentiation of individuals, ran through many of the lectures. Initially, discussion centered on the question, to what extent is the egg cell already organized in its earliest stages? Is there something brought to the egg by heredity, something to some extent predelineated? Or does form and heterogeneity emerge only gradually or epigenetically in the course of time? All of these discussions directly impinge on the more general debates about preformation and epigenesis. Those debates, revived on the continent in the 1880s and 1890s, also found expression at the MBL.[47]

To understand the debates, the modern reader must recognize that American biologists in the 1890s regarded preformation and epigenesis as closely associated with heredity and development, respectively. But they did not simply identify preformation with heredity and epigenesis with development. Rather, they saw both heredity and development as more complex. They did not even neatly distinguish heredity from development as biologists generally do now. Heredity did *not* generally mean transmission of characteristics or packets of information, after which development took over. Instead, heredity concerned whatever was passed from parent to offspring, and was regarded as a morphological phenomenon. Development of the individual involved its continuous response to the surrounding environment and acted as a physiological process with a morphological basis. Development of species paralleled development of

47. Jane Maienschein, "Preformation or New Formation—Or Neither or Both?" in T. J. Horder, J. A. Witkowsky, and C. C. Wylie, eds., *A History of Embryology* (Cambridge: Cambridge University Press, 1986), discusses the debates more generally.

individuals. Hyatt, in his only lecture to the MBL, in 1899, pointed out that heredity and reproduction both involve "the production of like by like."[48] The Americans generally believed that heredity accounts for similarities of form, while ontogeny brings variations. Both heredity and development act throughout an individual's life, and neither traditional preformation nor epigenesis strictly accounts for individual differentiations and development.

August Weismann (1834–1914) provided the favorite point of attack for the Americans. He offered a neat, apparently preformationist, view of individual development, suggesting that hypothetical units (the biophores, ids, idants) transmitted physical substance and thereby characteristics to offspring and on into individual cells. Elaborated by Wilhelm Roux (1850–1924), and seen in part in the introduction to his *Archiv für Entwickelungsmechanik der Organismen,* translated by William Morton Wheeler and reprinted here, was the Weismann–Roux hypothesis of qualitative cell division. That hypothesis held that the initial pool of determinants was, in the course of development, divided into a mosaic of different cells. Each of these then developed and became differentiated autonomously, according to its internal materials. Throughout the 1890s, the Americans rejected Weismann particularly, and the Weismann–Roux hypothesis, as too simplistic, too speculative, inadequate, and ad hoc. Indeed, Mathews expressed the common view when he indicted such theories as a "scientific misdemeanor."[49]

In 1890 Henry Fairfield Osborn (1857–1935) addressed the MBL on "Evolution and Heredity" and pointed out the tendency of researchers to embrace incompatible ideas. Evolution seemed to progress not randomly but along certain lines, as the neo-Lamarckians had shown. Strictly fortuitous variations would encounter the problems of swamping and regression. Thus, some version of inheritance of acquired characteristics seemed indicated. Yet Weismann's theory of the continuity of the germ plasm rejected such a possibility. The evolutionary process requires some theory of inheritance, Osborn held; indeed, explaining the "how, why, and when of variations" would furnish a "crucial test for any heredity hypothesis." Osborn believed that the evidence proved that Weismann's

48. Iris Sandler and Laurence Sandler, "A Conceptual Ambiguity that Contributed to the Neglect of Mendel's Paper," *History and Philosophy of the Life Sciences* 7 (1985): 3–70.
49. Albert Prescott Mathews, "The Physiology of Secretion," 1899, p. 183.

preformationist hypothesis of germ plasm continuity was severely problematic and inconsistent with other facts, but he nevertheless urged
openmindedness and a "liberal and generous spirit of discussion."[50] How
far heredity acted and when individual developmental responses to environment took over remained open questions for Osborn.

In 1893 the University of Pennsylvania botanist William Powell Wilson
(1844–1927) insisted that both internal and external conditions cause
variations. Thus, variations arise in part because of environment and are
thereafter neatly reproduced in the offspring. Heredity and developmental plasticity in response to environmental conditions work together
to explain variation; hence, neither preformation, closely associated with
internal factors, nor epigenesis alone prevails.[51]

E. B. Wilson directly addressed "The Mosaic Theory of Development" in
his lecture of 1893 and also called explicitly for something of a compromise between traditional preformation and epigenesis. Embryologists had
moved beyond the old biogenetic law, which stated that an individual's
ontogeny recapitulates its phylogenetic past. Though no embryologist
would any longer hold that the individual actually exists as such in the
egg, recent events had brought a move for some toward a version of preformation. For these new preformationists, structural units collect in the
idioplasm of the germ cell to produce something of a microcosm of the
future organism, Wilson pointed out. Such a microcosm theory of particulate inheritance began with Darwin and had been "pushed to its uttermost logical limit by Weismann," in Wilson's opinion. The mosaic theory
of Weismann and Roux maintained that the causes of differentiation are
mechanical and lie within the egg itself, carried out through cell division.
Yet, "brilliantly elaborated and persuasively presented as they are, they do
not at present, I believe, carry conviction to the minds of most naturalists,
but arouse a feeling of scepticism and uncertainty; for the fine-spun
thread of theory leads us little by little into an unknown region, so remote
from the *terra firma* of observed fact that verification and disproof are
alike impossible." The facts of experimental embryology had dealt a death
blow to the mosaic theory, Wilson felt, and thus "we have found good
reason for the conclusion that the mosaic theory cannot, in its extreme
form, be maintained." Yet neither could he accept the extreme epigenetic

50. Henry Fairfield Osborn, "Evolution and Heredity," 1890, p. 141.
51. W. P. Wilson, "The Influence of External Conditions on Plant Life," 1893, p. 165.

views of Hans Driesch (1867–1941) and Oscar Hertwig (1849–1922). Instead, he concluded that ontogeny begins with an egg, which already has
some definite constitution and is thus not raw, homogeneous matter,
while each step thereafter depends also on the interaction of the parts of
the organism. Therefore, something of internal or apparently predelineated self-differentiation but also something of responsive, regulative,
and hence epigenetic adjustment directs ontogenetic development. Initial organization is transformed into new organization through cell division and influenced by cell interaction.[52] This compromise position
received further articulation with John Ryder's and Whitman's lectures of
1894 and achieved its greatest degree of clarity with William Morton
Wheeler's lecture of 1898 (reprinted here).

The 1894 series of lectures in which Whitman expressed his own views
about preformation and epigenesis appears as a coordinated effort to
compare different points of view, to air varying opinions. "Cross-fertiliza-
tion works rejuvenation in theories as in organisms," Whitman wrote in
his prefatory note to that volume (published in 1895). Attacking the excesses of the neo-epigenesists, he noted that "an epidemic of metaphysical physics seems to be in progress—a sort of *neo-epigenesis.*"[53] Clearly
Whitman had no sympathies with those moves toward epigenesis which
rejected the importance of heredity altogether. Though he did not explicitly say so by 1894, he may well have begun to regard such work as Loeb's
mechanistic studies of development as at times bordering on such excesses, despite their provocative creativity. By 1894 Loeb had begun to
publish his work on artificial stimulation of cell division and production
of multiple embryos, followed by similar experiments and arguments for
epigenesis by Morgan and Driesch, among others.[54] As Loeb had said in
his lecture of 1893 and elsewhere throughout his work, all life phenomena
are determined by chemical processes. Differences in growth result from
different amounts of energy or differences in resistance to energy, since
energy is used to overcome resistance and produce growth. Therefore,
differences in form can be explained in terms of the chemical differences
which account for differences in resistance to growth. The extent to which
the external conditions acting on the egg can alone explain development,

52. Edmund Beecher Wilson, "The Mosaic Theory of Development," 1893, pp. 3, 5, 9.
53. Whitman, "Prefatory Note," pp. iv–v.
54. For discussion of Loeb's work, see Philip Pauly, "Jacques Loeb and the Control of Life:
An Experimental Biologist in Germany and America, 1859–1924," Ph.D. diss., Johns Hopkins
University, 1980.

without appeal to ancestral evolutionary conditions, remained a very heated issue in that year.[55]

In his rejection of extreme epigenesis, Whitman did not endorse traditional preformationism either. Development begins with something, something which is a product of the historical past and is influenced by heredity. That something is neither of the traditional alternatives, however; it is neither strictly performed nor homogeneous. One must avoid being sucked into the Charybdis of extreme preformation as surely as one must steer clear of the Scylla of extreme epigenesis. The real question for the day, Whitman clarified, was no longer preformationism or epigenesis. Rather the modern biologist should ask: "How far is post-formation to be explained as the result of preformation, and how far as the result of external influences?"[56] The germ is already organized in some way and yet external conditions exert an influence. How do the two work together, with what differential emphasis on each?

The embryologist John Ryder (1852–1895), who had directed the United States Fish Commission Laboratory in Woods Hole in 1888, expressed a view with a different emphasis in his discussion of "A Dynamical Hypothesis of Inheritance" in 1894. There he baldly asserted that Weismann's and other preformationists' theories were simply wrong. One must dismiss all such "imaginary corpuscles" as Weismann's and look to the mechanics and dynamics of development. The focus on ids and such presumed determinants "time will show to have been about as profitable as sorting snow-flakes with a hot spoon."[57] The ids are simply passing shadows, effects rather than causes of anything. Instead, the molecular, chemical organization of the egg (which is inherited) and the reciprocal influence of cells on each other, with energy as the only motive force, will explain development and heredity. Variations occur, on this view, as molecular systems interact with systems in the environment. The initial egg is neither preformed nor perfectly isotropic, according to Ryder. Rather, epigenetic development begins with a dynamically organized egg since it has a "determinate ultramicroscopic molecular mechanism."[58] Epigenesis with dynamic determinism: Ryder's lecture reveals the unsettled state of discussion of heredity and development, of preformation and epigenesis.

55. Jacques Loeb, "On Some Facts and Principles of Physiological Morphology," 1893, esp. p. 54.
56. Whitman, "Evolution and Epigenesis," 1894, pp. 221, 223.
57. John Ryder, "A Dynamical Hypothesis of Inheritance," 1894, p. 25.
58. Ibid., pp. 33, 28, 51.

Wheeler, in his lecture of 1898, brought a considered perspective on the debates, explaining the persistence of both epigenetic and preformationist viewpoints in terms of personality differences. Some found it easier to pass from the simple to the complex, to envision complex form as emerging from relative homogeneity. "The physiologist, who deals with processes, who is ever mindful of the Heraclitean flux, inclines naturally to" epigenesis, Wheeler wrote. "On the other hand, he who readily idealizes and schematizes, whose mind is endowed with a certain artistic keenness, an appetite for forms and structures, and a tendency to make these forms final patterns, eternal molds, more permanent than the substance that is poured into them—such a one will find more difficulty in understanding *how* the homogeneous can become the heterogeneous."[59] This latter is the morphologist, the Platonic preformationist. It is not clear which will prove to have the viewpoint more in accordance with the facts. Instead, agreeing with E. B. Wilson and Whitman, Wheeler suggested that the eventual resolution would lie somewhere between the extremes:

> The pronounced "epigenecist" of to-day who postulates little or no predetermination in the germ must gird himself to perform Herculean labors in explaining how the complex heterogeneity of the adult organism can arise from chemical enzymes, while the pronounced "preformationist" of to-day is bound to elucidate the more elaborate morphological structure which he insists must be present in the germ. Both tendencies will find their correctives in investigation.[60]

The ongoing discussion of preformation and epigenesis pointed toward the related issues of heredity and development, internal self-differentiating factors and external environment, and the significance of cell division. Such issues also underlay much of the discussion at the MBL in the 1890s and thus provide a background for the public evening lecture series. A closer look at those lectures, year by year, will reveal shifts in the assumptions and in the focus of discussion against a background of concern about the extent to which individual development is conditioned by inherited, internal, and hence preformed factors or by internal physiological responses to changing environmental conditions.

59. William Morton Wheeler, "Caspar Friedrich Wolff and the *Theoria Generationis*," 1898, p. 282.
60. Ibid., p. 284.

LEADING PROBLEMS: CHANGES THROUGH THE 1890s

The *Biological Lectures* certainly did not settle all questions about the relative importance of environment or internal conditions, of development or heredity, of epigenesis or preformation. But they did set out various alternative views of the issues and helped to clarify the points of disagreement and possible avenues for resolving disagreements.

Intimately connected with the questions about what directs development was work on cell organization and cell cleavage, pursued through cell lineage study and through investigations of fertilization. Many of these studies were undertaken with the purpose of providing solid data to address the value of particular working hypotheses about the nature of development. Specifically, the concern about the extent to which the egg is already organized at an early stage and what makes it differentiate further, which had a central part in the epigenesis and preformation discussions, appeared here as well.

1890 In fact, developmental concerns really lay at the core of the MBL discussion, as is evident even from the lecture titles through the 1890s. But development included heredity and evolution, so excluded relatively little. The first year's lectures, in 1890, reveal a traditional interest in the German morphological program. Thus, E. B. Wilson addressed "Some Problems of Annelid Morphology," for example. Using the opportunity to consider some general morphological issues, Wilson pointed out that the primary question concerned the derivation of the vertebrates. Taking Darwin's theory of common descent as a "splendid working hypothesis," the morphologist then had two questions: first, "What is?" and second, "How came it to be?"[61] Every question of morphology is also a geneological or historical inquiry, therefore. Since we can no longer see the actual ancestral forms, we must appeal to lower invertebrates and look for similarities which may reveal the parental characteristics. Three such noteworthy similarities include metamerism (or the production of segments), apical or unidirectional growth (at one end only), and concrescence (or the union of two halves along the median line). The origins and significance of these similarities remain unclear, Wilson concluded after reviewing the evidence. Only concrescence even appeared to be under a "satisfactory working hypothesis." But biologists should not despair because of the lack of positive results, Wilson urged. The good scientist must also pursue unsolved problems of the deepest interest such as these.

61. E. B. Wilson, "Some Problems of Annelid Morphology," 1890, p. 54.

Indeed, the "present need is for new facts, not for new theories. When the facts are forthcoming, the theories will take care of themselves." Wilson was himself embarked in 1890 on a detailed cell lineage study of the annelid worm *Nereis* and hoped to obtain sufficient descriptive data to shed light on questions of homology, ancestral forms, and such problems of morphology.[62]

Similarly, in 1890 assistant professor of zoology at Clark University John Playfair McMurrich (1859–1939) cited the value of explaining the origin of the metazoa from lower organisms. Ernst Haeckel's (1834–1919) gastraea theory of 1872 stood as the standard, suggesting that a monerula gave rise to a cytula, then to a morula, planoea (blastular form), and gastraea, after which the metazoa branched off. Haeckel's theory suffered from errors, so that various alternatives had arisen, several of which McMurrich discussed. Clearly the gastraea had never even existed, he maintained.[63] The important point here is that McMurrich, like Wilson, was presenting a problem, discussing evidence and alternative theories, then concluding that what biology most needed was more data and more study, in this case of the early development of invertebrates. As with Wilson's work, McMurrich reveals a traditional morphological perspective with an additional emphasis on careful observation and study and rejection of clever, neat theories that do not accord with the facts.

Osborn, Edward Gardiner Gardiner (1854–1907), and Morgan also addressed problems typically within the morphological tradition in that volume of 1890, Osborn looking at the relation of heredity (and stability) to evolutionary change, Gardiner at theories of death, and Morgan at phylogenetic relationships of sea spiders. Morgan's lecture was the most traditionally morphological and reveals a side of his work as a budding biologist that historians have usually overlooked. The lecture by Shosaburo Watase (1862–1929), "On Caryokinesis," reflected his study with Whitman at Clark University, for there Whitman had addressed the problem of Oökinesis. Watase's interest in the cell and cell changes and movements through early development shows the direction toward which work at the MBL soon moved, encouraged by Whitman—namely, toward careful analysis of early developmental changes.

1891 and 1892 Lectures for these years never appeared in print, but the

---

62. Ibid., p. 78; Edmund Beecher Wilson, "The Cell-lineage of Nereis: A Contribution to the Cytogeny of the Annelid Body," *Journal of Morphology* 8 (1893): 579–638.

63. See Ernst Haeckel, "Gastraea-Theorie," *Jenaische Zeitschrift* 8 (1874). 1–55; John Playfair McMurrich, "The Gastraea Theory and Its Successors," 1890, pp. 97, 91, 106.

Annual Reports record contributions to the series, nevertheless. In fact, even in the published years some lectures were not included in the volumes. In some cases, archival materials show that lecturers simply did not submit papers. Other lectures did not cover topics relevant to the series, though they undoubtedly appealed to the public audience. Morse's discussions of China and Japan fall into this category, as did some of the more descriptive and general discussions of evolution. For 1891 and 1892 Whitman had no publisher; only for the 1893 series did Ginn and Co. take over publication of the series.

1893 The second published volume of lectures came only in 1893, then, three years later. Those three years were very important ones, for in the meantime Weismann, Roux, and Driesch had been at work articulating theories and advocating a new experimental approach to embryology; moreover, the various studies of one-half and other partial embryos had raised new problems. It was no longer necessary to argue for the value of studying embryology. And embryology did not bring with it all the old assumptions of the morphological tradition, for that tradition had itself evolved and developed. As E. B. Wilson wrote in his 1893 lecture, the last ten years had brought a "remarkable awakening of interest and change of opinion" in embryology, especially about the significance of cell cleavage. In this major lecture, Wilson moved beyond the traditional program in morphology to new concerns, though those remained essentially morphological.[64]

In that essay, Wilson rejected the germ layer theory which had dominated embryology, arguing that one must begin with cells to trace cell lineages rather than begin with germ layers. As discussed earlier, Wilson looked favorably on an epigenetic view of development and rejected the Weismann–Roux mosaic theory. He urged that though biologists were "still profoundly ignorant of the nature and causes of differentiation, and of its precise relation to cell-formation," that nonetheless it had become clear in the preceding few years that differentiation often coincides with cell boundaries and hence was "not entirely independent" of cell formation.[65] Careful study of cleavage, of fertilization, of all the early developmental changes in cells and their relations to differentiation appearing

64. Discussed in detail by Alice Levine Baxter, "Edmund Beecher Wilson and the Problem of Development: From the Germ Layer Theory to the Chromosome Theory of Inheritance," Ph.D. diss., Yale University, 1974; Baxter, "E. B. Wilson's 'Destruction' of the Germ-Layer Theory," *Isis* 68 (1977): 363–374.

65. Edmund Beecher Wilson, "The Mosaic Theory of Development," 1893, p. 14.

later were indicated as a priority for biology—all assuming that the cell was a, if not the, fundamental unit of development and heredity.

Several other papers in 1893 and later addressed precisely such phenomena. "Fertilization of the Ovum" by Edwin Grant Conklin (1863–1952) examined the relative roles of nucleus and cytoplasm for development and concluded that "the *entire* cell is still the ultimate independent unit of organic structure and function." Though the vast majority of biologists were coming to endorse the nucleus as the bearer of heredity, Conklin insisted that both nucleus and cytoplasm played crucial roles in both heredity and development. After all, fertilization brings together parts of both male and female cytoplasm as well as nucleus; for example, the sperm aster is cytoplasmic. The "independent unit of structure is still the entire cell, not cytoplasm alone, nor nucleus alone, but the two together," he maintained.[66] This unified, coordinated cell is where embryologists should begin their study of the patterns and causes of differentiation.

Watase agreed, also in 1893, that morphologists were explaining origin and development in terms of cell growth, and physiology in terms of component cells. The cell may well have arisen by a symbiotic union of two units, becoming the cellular nucleus and cytoplasm. The symbiotic relations are adaptive so that now nucleus and cytoplasm need each other, with neither any longer capable of survival alone and with each keeping the other under control. Fertilization simply brings the synthetic product of one cell from two, and cell division is essentially incidental to increase in both nucleus and cytoplasm. Thus, the symbiotic cell retains a primary biological importance for Watase.

Yet Whitman objected to what he saw as "the inadequacy of the cell theory of development," which Conklin, Watase, and others accepted. Cells are important, of course; indeed, the organism is the product of cell formation. But above the cell, the whole organism predominates. We see a similar position expressed by Whitman's colleague at the University of Chicago, Charles Manning Child (1869–1954), in his 1899 lecture. "It is the *organism—the individual, which is the unit and not the cell,*" Child emphasized.[67] Cells act together, as a unit, because of the organization of the whole. Differentiation and formation of the organism are not due to cell

66. Edwin Grant Conklin, "The Fertilization of the Ovum," 1893, pp. 34, 32.
67. Charles Manning Child, "The Significance of the Spiral Type of Cleavage and Its Relation to the Process of Differentation," 1899, p. 265.

division, to cleavage, or to some mysterious "cellular interaction," Whitman agreed with Child. For the claim that the egg already experiences definite organization, he believed he had decisive proof. All cell divisions do not simply divide up the differentiated areas already laid out but can cut across those areas. Whitman agreed with Huxley that cells are like sea shells; they are the effects that show where the tides, or developmental processes, have acted. They are only effects, so that cell division cannot be a cause of differentiation. Cell lineage, for Whitman, was valuable for demonstrating the patterns by which the whole functioning organism gained its organization. As Lillie wrote of Whitman after his death,

> Whitman took a strong and independent position, basing his conclusions not merely on comparative embryology but also upon the comparison of protozoa and metazoa. He protested against the view that organization is the product of cell formation, and insisted that "organization precedes cell formation and regulates it." He contrasted the Cell-doctrine with what might be called the Organism-doctrine. He insisted that, "an organism is an organism from the egg onward, quite independent of the number of cells present," and that cleavage is not a process by which organization arises, but that organization precedes cleavage."[68]

Despite his strong view, Whitman did not force his ideas on others. Though dogmatic about principle, one of his principles was to remain scientifically undogmatic and to support and encourage proper scientific research. Thus, he encouraged open discussion and investigation, even when he disagreed with the conclusion. As long as the investigation was well designed and the results solid, and as long as the conclusions were expressed so as to reveal the open questions or points of disagreement, he did not object.

Whitman supported Loeb, for example, even when Loeb argued for a strictly mechanical, epigenetic view of development which seemed to degrade the significance of heredity. Loeb rejected many of Whitman's conclusions, yet Whitman supported him at the University of Chicago and made efforts, eventually unsuccessful, with the administration there to

---

68. Whitman, "The Inadequacy of the Cell Theory of Development," 1893, p. 124; Lillie, "Whitman," p. xliv.

ensure Loeb's staying. Whitman also invited Loeb to the MBL to head the physiology program which he regarded as a neccesary complement for morphological work.[69] By 1894 Loeb had embarked on his physiological program for morphology, as he considered it. He sought to determine the causes, inevitably mechanical and chemical, of animal forms. With a series of experiments on artificial production of cellular or of nuclear division, he established features of the mechanics of growth and formation which emphasized physicochemical tropisms directing development. "Heredity" offers no explanation of that fundamental question, what causes the arrangement of the different germ regions, Loeb insisted. Thus, neither parental conditioins nor other evolutionary factors play any role. Instead, all phenomena are determined by chemical processes, with differences in energy amounts or resistance explaining differential growth. Physiological morphology will provide the laws of organization due to chemical activity of the cell. But it also has the synthetic goal of allowing man to form new combinations from nature. Control as well as analytical understanding was what Loeb sought.[70] Such a program would not have appealed to Whitman, or to most of the others at the MBL by 1894, but Whitman accepted Loeb's alternative "standpoint" as important for the cooperative idea of biology.

In a different way, we find tolerance of occasional lectures by botanists. After all, the MBL considered all of biology, in theory at least. Difficulties in obtaining a botanist with interests parallel to Whitman's and thus in integrating botany with zoology, and the less "sexy" nature of botanical work at the MBL contributed to botany's secondary status there as elsewhere in leading American biology programs.

1894 In two series of lectures, 1894 and 1898, Whitman encouraged consideration of the details and significance of cell division. The 1894 lectures he regarded as dealing "with one or another side of the problem of organic development—that problem which has led, and which will most likely ever continue to lead, the biological sciences."[71] Comparison of different standpoints could prove valuable to all sides. The published vol-

69. MBL *Annual Reports*, plus various items in the University of Chicago Archives in the Whitman Papers, Presidential Papers, and Zoology Department records.

70. Jacques Loeb, "On Some Facts and Principles of Physiological Morphology," 1893, pp. 37, 53; Loeb, *Untersuchungen zur physiologischen Morphologie der Thiere* (Würzburg: Hertz, 1891; 1892) I, Heteromorphosis. II. Organbildung und Wachsthum; Pauly, "Jacques Loeb."

71. Whitman, "Prefatory Note," 1894, pp. iii–ix.

ume began with several general lectures addressing problems of development, meaning ontogeny, heredity, and evolution, then focused more directly on cell action.

The Tufts University physicist Amos Emerson Dolbear (1837–1910) and John Ryder opened the series with theoretical examinations of heredity from a largely mechanistic, physicalist perspective. Then Osborn, in a lecture reprinted here, cited the need for using inductive approaches in biology, for rejecting excessive speculation, and for avoiding the "unnatural divorce" of specialties from their common goals. In particular, embryologists and paleontologists should work together to develop a theory of both these intimately related phenomena or their fields of study.

An essay by George Baur (1859–1898) also considered evolution, but more directly. In fact, his discussion of speciation in the Galapagos Islands represents a notable departure from most of the published lectures in the series. Though a number of oral presentations had considered such traditional evolutionary questions, very few appeared in the published papers. Born in Germany, Baur had served as an assistant at Yale, then went to Clark University in 1890, as Whitman had. After a visit to the Galapagos in 1891, Baur moved with Whitman to Chicago in 1892, as assistant professor of comparative osteology and paleontology. Whitman knew that Baur had put forth a theory, based on his own study of Galapagos species, which directly contradicted Darwin's and which had stimulated considerable controversy.[72] Presumably that explains Whitman's invitation to Baur to serve as instructor at the MBL in 1894 and to present an evening lecture. There, Baur expressed his view that the harmony of species distribution could only be explained if the islands had resulted from subsidence. Instead of the islands' having arisen from the ocean and having been populated by migration from the mainland as Darwin suggested had occurred, they were isolated by subsidence, or a rise of water level. Isolation produced speciation, he argued. Only that theory could explain the harmony evident among species on the different islands, Baur maintained even in the face of powerful opposition. As he began to gain supporters, his theories received considerable attention.

Returning to a more familiar theme of the published lectures, E. B. Wilson urged the value of embryology, and indeed of morphological embryology. The embryological method based on the biogenetic law, which assumed that embryological development (ontogeny) repeated the an-

72. William Morton Wheeler, "George Baur," *American Naturalist* 33 (1899): 15–30.

cestral history (phylogeny), had been basic to morphology. But both that embryological method and morphology urgently needed revision, since both had come to be regarded with distaste. "The truth is," Wilson admitted, "that the search after suggestive working hypotheses in embryological morphology had too often led to a wild speculation unworthy of the name of science." Nonetheless, embryology can be worthwhile, and even valuable, as a guide to homologies. Researchers needed a "trustworthy basis of interpretation," a candidate for which Wilson went on to suggest. Biologists should not abandon embryology altogether, nor the traditional descriptive and comparative accumulation of facts which are, after all, the "very framework of biological science." But avoiding speculative excesses does not also require a complete endorsement of experimental methods and rejection of more traditional methods. Embryology must use various methods and provide itself with a solid foundation, recognizing that "the greatest fault of embryology has been the tendency to explain any and every operation of development as merely the result of 'inheritance,' overlooking the vital point that every such operation must have some physiological meaning for the individual development, hard though it may be to discover."[73]

Turning directly to the significance of cell division in his essay, John McMurrich also acknowledged the recent remarkable developments in embryology, especially with cell lineage and physiological studies of the embryo. Researchers had also turned attention to the particular question of what determines the direction of cell cleavage. Because it proved difficult to discover adequate mechanical causes, the experimental physiologists had tended to look outside the cell itself, to external causes such as pressure and gravity. This led to apparent but possibly deceptive simplicity. Probably none of the extreme views would prove correct, McMurrich thought, in keeping with the MBL tendency to adopt intermediate positions.

Watase examined the origin of the centrosome, concluding that it came from cytoplasm and that the cytoplasm thus had a "certain endowment" not fully recognized. Since the centrosome seemed associated with protoplasmic movements, its role in directing cell division and cell action should prove important.

In 1894, then, Whitman had organized the series of lectures to attack questions about morphology and about the role of embryology with respect to heredity and evolution, and to raise considerations about the role

73. E. B. Wilson, "The Embryological Criterion of Homology," 1894, pp. 103–104, 123–124.

of the cell and cell division. These were all central issues of the day for biologists and were not particular pet problems of Whitman's. They all arose from the flurry of experimentation and interpretation pouring out of Germany, against the background of cell lineage and other morphological work carried on in the first years at the MBL. Wheeler's translation of the introduction to Wilhelm Roux's new *Archiv für Entwickelungsmechanik* illustrated the interest of Americans in—though not their acceptance of—the German work. That important essay reveals Roux's self-confident, indeed bombastic, style and his commitment to a mechanical and experimental program for embryology, which provided such a center of interest for MBL researchers.

1895   The year 1895 brought a move to consider general biological phenomena. The seemingly assorted set of papers actually represents a cross section of current, largely nondevelopmental concerns. Infection, immunity, Huxley, paleontology, descriptions of segmentation, bibliographic materials, transformation of plant parts: these represent biological subjects, broadly interpreted, and may well have reflected Whitman's attempt to remind the MBL audience that biologists should be concerned with more than development, heredity, and evolution. The papers by Dolbear and Watase bear particular interest in revealing those general concerns.

Dolbear presented a view of biological knowledge based on mechanical, chemical, and physical causes. Yet he, like Whitman, did not adopt a radical reductionist position, in which all life is regarded as explainable in terms of minute particles acting just as they would in any physical system. Such may be the actual basis of life, but life exhibits a level of complexity which may warrant explanation in terms of material units and their relations. Whitman suggested that, just as there is an organic chemistry, there might well be an organic physics, distinct not in kind but in complexity. A mechanist rather than a vitalist, Whitman nonetheless regarded life as calling for explanations falling short of extreme, radical mechanism. He embraced what might today be considered a form of emergentism. As Whitman had said in 1894:

> The search for ultimate units of organization in the egg—that is, smallest elements capable of organic growth and self-division— has already led directly to the discovery of *mechanism,* where molecular epigenetics had disputed it. The molecule is no doubt universal and very mighty, only perhaps not quite almighty. It is quite conceivable that there should be something at least as far

above the molecule as the molecule is above the atom. Indeed, there seems to be a considerable number of units actually visible in the cell, which are certainly quite as real as the molecule, and which differ from it in having those fundamental attributes of growth and self-division which appear to be peculiar to every grade of organic life. Every such unit may be reducible by chemical disintegration to molecules, but we should hardly accept that as proof that no organization above molecules preceded the dissolution. There is no warrant for the assertion that life is something different from, and independent of, matter and energy. That is the mistake of vitalism. On the other hand, there is no warrant in decomposition for identifying dead mechanism with living mechanism.[74]

Whitman clearly found it important to assess the proper basis for biological study and the relation of biology to physics.

In the lecture series of 1894, Dolbear had discussed "Life from a Physical Standpoint." There he had stressed the necessity of hypotheses for guidance in science and had cited Darwin's theory of natural selection as the "only rational hypothesis" of the time concerning the "nature of life." The best hypothesis about the composition of organic nature was also that phenomena of life are resolvable into physical and chemical processes, so that life is an attribute of matter just as surely as electricity and gravitation are. Essentially, for Dolbear, life is the result of a field of activity resulting from a complex of vortex atoms in the ether, again just as gravitation, magnetism, electricity, or chemical attraction are. Yet he acknowledged that there could be something different for organic matter, though certainly not a vital force or vital entity. Dolbear's particular view of physically based life was very like Whitman's own, though worked out much more clearly and in terms of current ether and field physics of the 1890s.

Whitman invited Dolbear back for an additional series of five lectures in 1895, two of which he published. In the first, Dolbear denied that any such thing as a vital force exists; indeed, energy itself is a product of more basic changes. Thus, a proper explanation, in biology as elsewhere, "is complete when the physical and chemical antecedents have been presented in their order and quantitative relations." Life may be more complex than, but not fundamentally different from, nonlife. In his second lecture, Dolbear reiterated the essential chemical and physical nature of

74. Whitman, "Prefatory Note," p. vi.

living matter, then went on to deny any form of mind–body dualism. There simply is no evidence that mind exists separate from matter. Rather, consciousness results from nervous energy, itself a product of motion of ultimate particles of matter. Mind results just as magnetized iron results, due to rearrangement of molecules.[75]

Certainly not all biologists in the world would have agreed with such physicalist but not radically reductionist proclamations, but generally the American researchers, and especially those at the MBL, did. Even those sympathetic to Driesch's later turn toward philosophy and his rejection of the Weismann–Roux form of predeterminism and self-directed developmental mechanics did not embrace vitalism. Dolbear's clarifications and explicit statements undoubtedly helped many who had never really articulated their assumptions about the nature of the phenomena they investigated.

Following Dolbear's contributions came a lecture by Watase, this time explicitly examining the physical nature of a peculiar biological phenomenon. Animal phosphorescence might well seem like a peculiar living phenomenon, as indeed it had in the past. Yet, as Watase urged, we can know life only through its physical, chemical, and mechanical manifestations, of which the emission of light is one example. Light production is of no use to some insects but is of value to others. The ability to produce light has what might appear to be two causes, one proximate, the other ultimate: "While the production of light may be regarded as belonging to the same ultimate cause as that of heat, the proximate cause of the luminosity in the animal kingdom may be due to a variety of secondary circumstances."[76] In fact, it is the background or ultimate substructure which really explains the phenomenon. Chemically, production of light requires oxygen, hence respiration. Perhaps the oxygen combines with some dead substance prepared by the cell normally and in some cases produces luminescence. This one phenomenon may well be connected closely with the "whole mystery of life."[77] With Watase's lecture, then, we see a particular biological application of Dolbear's and Whitman's general viewpoint concerning the chemical and mechanical nature of life.

Paleontology is a morphological discipline, asserted William Berryman Scott (1858–1947) in his lecture. Paleontology considers "the factors of

75. Amos Dolbear, "Explanations, or How Phenomena Are Interpreted," 1895, p. 73; Dolbear, "Known Relations between Mind and Matter," 1895, p. 93.
76. Shosaburo Watase, "On the Physical Basis of Animal Phosphorescence," 1895, pp. 107, 113.
77. Ibid., p. 118.

evolution, the causes which determine the development of new forms, and the problems of heredity which are inseparably connected with them." Ultimately, the decisive evidence will probably come not from paleontology itself, however, but from "the physiological and experimental method," meaning experimental embryology. Indeed, "experimental embryology has already won some notable triumphs, and that is a physiological quite as much as a morphological province."[78] Cooperation of different disciplines and of different approaches will prove the only way to cut through the vast accumulation of facts to answer the pressing questions, Scott concluded, echoing a favorite theme at the MBL.

1896-1897 A selection of the lectures from both 1896 and 1897 appeared together as the fifth volume of the series, perhaps partly because of Whitman's struggles with the Trustees over finances in 1897 and the resulting near failure of the MBL to open at all that year.[79] Again, we find a mixed set covering the range of topics that Whitman regarded as properly biological. Lectures on the sparrow, on paleontological methods, excretion, neural terms, plant classification, and biological stations represent the range of topics. All address subjects of general interest and tie the discussion in with broad concerns. Four of the essays focus more directly on cell actions in development, amplifying a favorite theme at the MBL.

Conklin expressed the most widespread sentiment at the MBL with his emphasis on embryological questions: "Philosophically, the most important problems of biology are those which concern the origin of a new individual, the genesis of a living organism ... The mystery which hangs about the process of progressive and coordinated differentiation by which the egg cell is transformed into the adult never loses its charm nor ceases to be a mystery." His insistence on the value of both observation and experimentation for embryological work, in opposition to Roux's demand for experimentation alone, also reflects an attitude typical of MBL scientists. As Conklin stressed, "There is no such sharp distinction between observation and experimentation in biology as is sometimes assumed; neither method can arrogate to itself a monopoly of certitude regarding facts or causes."[80]

78. MBL *Annual Reports* (1897); discussed in Lillie's history of MBL in a number of places and in letters of Whitman to Conklin, Lillie, Wilson, and Morgan, MBL Archives.

79. William Berryman Scott, "Paleontology as a Morphological Discipline," 1895, pp. 58, 60.

80. Edwin Grant Conklin, "Cleavage and Differentiation," 1896–1897, pp. 17–18.

Evidence was accumulating, Conklin said, that something like Wilhelm His's organ-forming germ regions might in fact exist. Assuming they do, that the egg is thus already organized to some degree, then what significance does cell division have for differentiation of the organism, that "most important problem of biology?" Cleavage may logically either (1) follow the pre-established regions; (2) cut across those regions; or (3) merely chop up homogeneous material (if one rejects His's organ-forming germ regions). The first of these Roux captured with his mosaic theory; the second was Whitman's "organization theory"; while the third was the "homogeneity theory" endorsed by Driesch and other extreme epigenetic physiologists. In fact what happens, Conklin asserted, is that all three pertain, but for different organisms. Some, such as annelids and gastropods, have a very regular, determinate, or mosaic, cleavage. Echinoderms, vertebrates, and others generally experience indeterminate cleavage, whether explained by (2) or (3). Experimental production of whole embryos from half the material does *not* support the contentions of Driesch in particular, that the egg material never had any organization, that cleavage was thus proven indeterminate. Rather, the artificial experimental case showed nothing more about normal development than did regeneration studies. Normal development might proceed in a highly constant and determinate manner, but disruption might bring other factors into play and produce necessarily adjusted and hence indeterminate cleavage.

Determinate cleavage under normal conditions results not from chance nor from extrinsic mechanical factors, which direct indeterminate cleavage. Rather, the cause must lie with the protoplasmic structure of the cells. Constancy of the first cleavage indicates constant protoplasmic arrangements in the unsegmented egg, or intrinsic factors. All of these intrinsic factors depend on the organism's history, for "the reason that a certain blastomere arises in a certain way, passes through a definite developmental history, and in the end gives rise to a definite part is at bottom the same reason that the egg of a given animal passes through a definite history and gives rise to a definite organism."[81] Similarities, or constant mosaic cleavages, generally predominate in the earliest stages, while divergence of cell division and differentiation increase as development progresses. Conklin's distinctions and clarifications became a standard, even for those who disagreed with his conclusions, for subsequent discussion of cell cleavage.

81. Ibid., p. 34.

This essay also serves to invalidate the incorrect impression that some modern researchers have of the work at the MBL. J. M. W. Slack wrote recently in a review of William Jeffrey's and Rudolf Raff's edited volume of MBL lectures from 1983, that

> ever since the days of Whitman, Wilson, and Conklin, Woods Hole embryologists have been of the school that believes that regional specification arises from the passive partition by the cleavage planes of regulatory molecules differentially localized within the egg. Such mechanisms do indeed operate in some cases, and the second half of the book includes much of the best evidence. However, it seems to me that in all these cases the localization only represents the *first* decision and that all the remaining pattern of the embryo becomes specified as a result of interactions between the parts.[82]

Conklin, Whitman, and Wilson would surely have agreed. Such a statement fails to recognize the sophistication of the MBL work of the 1890s and the understanding by the principals that many features, indeed more than Slack might well admit, *do* work together to effect embryonic development.

Katherine Foot (1852–1944?) presented the first of two lectures at the MBL given by women. Her consideration of the origin and function of the centrosome in development follows the standards of the time. She concluded that two centrosomes exist, one from each parent, and that both are cytoplasmic rather than nuclear elements. Also typical of the time, she presented her conclusions as preliminary, as the best available hypothesis, to be revised in the face of new data and new ideas. Whether because of her own reportedly somewhat negative personality, or because of feelings of insecurity in her role as first woman lecturer, or because of an exaggerated sense of honesty, she concluded, excessively apologetically, that "I believe I really *know* very little about the subject, and when I have read more I shall probably know less."[83]

Henry Crampton's (1875–1956) lecture on coalescence experiments extends Gustav Born's (1851–1900) transplantation and grafting experiments to the Lepidoptera. Just as Conklin and Albert Davis Mead (1869–1946)

---

82. J. M. W. Slack, review of William Jeffrey and Rudolf Raff, eds., *Time, Space, and Pattern in Embryonic Development*, in *American Scientist* 73 (1985): 293–294.

83. Katherine Foot, "The Centrosomes of the Fertilized Egg of Allolobophora Foetida," 1896–1897, p. 57.

considered cell division as "close to" the most fundamental of biological problems, Crampton regarded animal grafting and resulting coalescence or tissue hybridization as "among the most important."[84] He hoped to gain further information about the role of heredity in producing color by grafting together pieces of differently colored species. Did the gonad and color have a common cause or did the gonad (and hence heredity) cause color? He succeeded in grafting together pieces of tissue, notably fronts and backs of different individuals. Unfortunately, the results concerning color production remained inconclusive, and he could not answer his own question.

Whitman evidently acknowledged the grafting techniques and possible results as valuable for embryological research. He tried on several occasions to persuade Ross Granville Harrison (1870–1959), a Trustee who spent some time at the MBL during several summers and who was a close friend of several regulars there, including Conklin and Morgan, to discuss his transplantation work in frog embryos. Not one to enjoy giving public lectures, Harrison came close to agreeing but never actually presented a lecture at the MBL.[85]

1898 A series of coordinated and largely embryological lectures came in 1898, the year after Whitman's major battles with the Trustees. He still hoped to gain a permanent endowment for the lab and thus to make it an independent, national biological center with a secure financial base. To that end, he still sought publicity where he could find it. The *Biological Lectures* had proven successful in gaining wide and positive attention.[86] Perhaps he hoped that a series of lectures by his leading regular researchers and several students from Chicago on a topic of central importance in American work would exhibit the advances made by Americans and would reveal a productive program of research. Thus, most of the lectures focus on factors in differentiation. Two lectures on evolution by MBL regulars, former assistant director of the MBL Herman Carey Bumpus (1862–1943) and the paleontologist Scott, filled out the traditional topics. In addition, 1898 witnessed the introduction of Whitman's work on animal behavior, which had come to dominate his own research interests even while he still guided his students into studies of cell lineage

84. Henry Crampton, "Coalescence Experiments upon the Lepidoptera," 1896–1897, p. 219.

85. Ross Granville Harrison Collection, Yale University Manuscripts and Archives, and Whitman Collection, MBL Archives.

86. Whitman, "Prefatory Note," p. iii.

and differentiation. W. W. Norman's paper on pain illustrates a similar concern.

E. B. Wilson's two papers are both classics, revealing his clear thinking and solid morphological viewpoint, stimulated by his work at John Hopkins under Brooks and at the MBL with Whitman. These papers, together with his earlier contributions, nicely summarize his general assumptions in biology. The first acknowledges that the promise that knowledge of structure could provide sufficient information to understand the physiology of development was fading. That hope "is giving way to a conviction that the way to progress lies rather in an appeal to the ultra-microscopical protoplasmic organization and to the chemical processes through which this is expressed. Nevertheless, it is of very great importance to arrive at definite conclusions regarding the visible morphology of protoplasm."[87] As Darwin had suggested, protoplasm consists of heterogeneous corpuscles, an intricate network of different substances. Darwin's hypothesis of pangenes will not likely prove correct, nor are we likely to find a perfect alternative to account for physiological changes since the protoplasm is extremely complex. The structures we see are probably secondary, with finer and finer structures in the background so that eventually we reach what appears to be a homogeneous substance in which the structure remains invisible.[88]

Understanding the protoplasm is necessary but not likely to be fully realized: that dilemma ultimately contributed to the move by Wilson and others away from the research emphasis of the 1890s. They moved from cells and their morphological characters to different, more immediately productive research ventures, notably in cytology of the chromosomes and genetics.

Wilson's second paper is at first sight a throwback to his earlier cell lineage studies of 1891–1892. Then he had worked on the cell lineage of *Nereis* and discovered that Conklin, then at the United States Fish Commission at the Johns Hopkins table, was undertaking a similar "cell-counting" study of the gastropod *Crepidula*. Wilson walked over to meet Conklin; the two discovered striking similarities in the manner of development of their two forms. Wilson introduced Conklin to Whitman, who offered to publish Conklin's results in his *Journal of Morphology* and also thereafter introduced more and more students to cell lineage work. The

87. E. B. Wilson, "The Structure of Protoplasm," 1898, p. 2.
88. Ibid., pp. 15–16.

ideal of a group of individuals specializing on different organisms, then working cooperatively to arrive at generalizations by comparing their results and interpretations, obviously appealed to Whitman.[89]

Lillie recalled in his unpublished informal autobiography his first introduction to Whitman after he had determined to attended Clark University in order to work with Whitman. Whitman promptly encouraged Lillie to spend the summer at Woods Hole, and there set him to work tracing the cell lineage of the freshwater oyster *Unio*. This meant that Lillie had to trek back and forth by train from Woods Hole to a pond in Falmouth, lugging his collecting materials with him.[90] In the early years of the MBL, Whitman and his students and associates took cell lineage seriously.

Wilson's second paper of 1898 serves as an apology for that work at a time when the lure of experimentation and physiological questions had become strong if not irresistible. Cell lineage work and the embryological and evolutionary questions it illuminated remain at the core of biology, Wilson maintained. Every individual presents us with two problems together, he suggested. While it is a complicated mechanism in itself, maintaining a complex adaptive equilibrium of its own parts and with its environment, each individual also represents an adaptive product of past conditions. The physiologist or morphologist therefore has two tasks, the second including the study of that historical background. Cell lineage provides a new embryological method but provides no "open sesame" of perfect answers. Yet the suggestions and definite results are useful to demonstrate probable ancestral conditions. Unquestionably, ancestral reminiscences occur in even the earliest developmental stages and they serve as guides to genetic affinities just as they raise suggestive questions in pure morphology.[91]

Lillie followed Wilson with a general summary of the significance of his work on *Unio*, for cleavage in particular. Cleavage, when determinate as it is in *Unio*, reveals definite adaptations, results of internal conditions. Perhaps not the direct product of organization of the egg, the cleavage nonetheless follows some definite orientation or intercellular processes. Cell lineage work reveals the patterns of cleavage and exhibits a strong deter-

89. See Jane Maienschein, "Cell Lineage, Ancestral Reminiscence, and the Biogenetic Law," *Journal of the History of Biology* 11 (1978): 129–158.

90. Frank Rattray Lillie, "Autobiography," section 6, unpublished manuscript, MBL Archives, pp. 27–28; Lillie, "The Embryology of Unionadae," *Journal of Morphology* 10 (1895): 1–100.

91. E. B. Wilson, "Cell-Lineage and Ancestral Reminiscence," 1898, pp. 24, 40.

minate character, unchanged by the external factors which some had argued were alone decisive.

Conklin looked at the causes of differentiation, also illuminated by cell lineage work since he regarded the most important phenomena of development as those earliest stages: the development of polarity in the egg, for example. The causes are said to be due to the protoplasmic structure of the germ, but that means little without knowledge of the intermediate steps, he argued. This weakness of any hereditarian program, including genetics, continued to disturb Conklin, the committed epigenesist. Egg division and processes such as maturation, fertilization, and cleavage seemed the product of mechanical movements, perhaps vortex movements of the protoplasm. Here as elsewhere "the cell acts as a whole, and in the interaction of its various parts are to be found the causes of all vital phenomena."[92] Though he had not established the causes of the intermediate steps between egg origin and fully formed and differentiated individual, he was convinced that careful descriptive study of the protoplasm would reveal changes at least contributing to an explanation of cell division. Like Wilson and Lillie, Conklin reaffirmed the value of the morphological study of early cell actions. Also like Wilson and Lillie, he recognized that such changes reflect adaptations to past as well as present and later developmental conditions. Whitman's students Aaron Louis Treadwell (1866-1947) and Albert Davis Mead pursued similar lines of research and reasoning.

Cornelia Clapp's paper focused once again on the significance of the first cleavage plane. Instead of asking whether it resulted in a regular and definite way or whether it cut across differentiated regions, however, she looked to a slightly later stage of development. She asked whether the first cleavage plane decides the direction of the embryonic axis, as Roux and Eduard Pflüger (1829-1910) insisted, or whether that axis is already set in the egg, or whether the axis becomes established later. Whitman had encouraged Clapp to pursue her own research and had suggested that she examine the toadfish *Batrachus*, which has an adhesive disk to hold it in a fixed position. This unusual feature proved extremely valuable in allowing her to rotate the dish to which the egg attached itself and to test the resulting changes. Her preliminary results, published earlier, had demonstrated that the first cleavage plane and the embryonic axis coincided in only three of twenty-three cases, suggesting that some fac-

92. E. G. Conklin, "Protoplasmic Movement as a Factor of Differentiation," 1898, p. 90.

tor other than cleavage set the axis. Roux and Pflüger had challenged those results, with the effect that Clapp redoubled her efforts to establish her point.

In fact, she retorted, their work with frogs required fixing and marking procedures "which must always cast some doubt on the reliability of the results."[93] Her studies showed without question that the first cleavage does not coincide with the axis. The axis appeared in different cases in all directions from the first cleavage, a fact that she established with unfailing logic and clear evidence. Thus, "the study of cleavage has not yet given us any such fundamental laws of development as the mosaic theory claims." And "the opinion is gaining ground that the phenomena of cleavage are to be regarded as the expression of non-differential rather than qualitative divisions of the germinal material."[94] In thus challenging the strongly stated conclusions of Roux and his compatriots, Clapp had full support of the program in cell lineage and other careful descriptive and comparative morphological work encouraged by Whitman at the MBL. The physiological studies of Loeb and others, which emphasized the importance of physiological regulative responses to external conditions, also lent credibility to Clapp's attack on the simplicity of Roux's mechanically self-differentiating system.

Ironically, just as the set of lectures of 1898 summarizing the value of cell lineage work and the way in which heredity worked (for Loeb) and physical conditions shape development in regeneration (for Morgan) were delivered—and just as the group had come more closely together then it had previously in its series of summer lectures—Whitman turned his attention elsewhere. He had become attacted to a different set of biological problems altogether, to animal behavior. In some ways, his work signaled the coming of greater specialization and the loss of that ideal union. It also signaled the beginning of Whitman's withdrawal from MBL research and administration. But it represents Whitman's best work, and for that reason the paper of 1898 is reprinted here.

Psychogenesis, or the study of the emergence of habits, instincts, and intelligence, was as much a problem of natural history as Whitman's earlier work had been. Instinct, like development, resulted from adaptations chosen by natural selection. Indifferent organic material serves as a foun-

---

93. Cornelia Clapp, "Relation of the Axis of the Embryo to the First Cleavage Plane," 1898, p. 144.
94. Ibid., p. 151.

dation, from which gradual modifications bring instinct, then intelligence. "The adaptation of acts to purposeful ends must not be accepted too quickly as proof of intelligence in the doer," for adaptation may result from slow and blind selection among alternatives. From the leech *Clepsine* to his pigeons, Whitman examined details of behavior and suggested how they can be seen as results of selection and not of prior intelligence by the possessor. Choice comes when increased plasticity invites greater interaction of stimuli and hence facilitates conflicting alternative impulses. Education and learning result. "Plasticity of instinct is not intelligence, but it is the open door through which the great educator, experience, comes in and works every wonder of intelligence."[95] Development of intelligence and learning, then, is like development of form in that the individual begins with something inherited and adapted to past conditions, then learns through experience to respond to present conditions.

After 1898 Whitman continued to spend his summers at the MBL, but he became increasingly less a part of the research there. Indeed, by 1898 Whitman had been deeply discouraged by the conflicts of the previous year and had begun to withdraw from the MBL, from the University of Chicago, and from leadership roles generally. Though always available as an advisor, he no longer stimulated students to undertake projects closely related to his own. Most of the leading MBL researchers, such as Wilson, Conklin, Loeb, and Morgan, continued along their own paths of research, pursuing problems which grew out of their interests of the 1890s. Whitman's work diverged from theirs, though it also followed along lines initially included as part of the broad morphological tradition. Instead of studying marine organisms or worrying about the nature of biological research laboratories or American biology in particular, he turned to his pigeons. He laboriously packed them and hauled them back and forth from Chicago to Woods Hole for a few summers of research.[96] But the sort of experimental work such as Loeb's and Morgan's which was gaining predominance at the MBL, work which promised definite and relatively quick

95. Whitman, "Animal Behavior," 1898, pp. 298, 336, 338.

96. Ishikawa, "Whitman," 1911, p. 18: "His pet pigeons were abundant in numbers. He told me that when he started on a trip to the east, he took the pets with him. Year after year, the number is multiplied and he experienced increasing trouble in the transportation of the pets. So he quitted to come to Woods Hole, assigning the directorship of the Marine Biological Laboratory to Professor Lillie."

results but often without full consideration of the evolutionary historical past, did not appeal to Whitman.

1899 The last series of lectures, in 1899, reflects Whitman's own evolving concerns and his estrangement from the experimental work of embryology and physiology. Lectures on evolution and behavior dominate that year's offerings, though papers by his University of Chicago colleagues Mathews and Child discussed more traditional problems, as did Morgan's second installment on regeneration and Loeb's on fertilization.

Morgan's and Loeb's lectures, reprinted here, both reflect a move away from evolution and heredity which was seen as taking place in physiological work throughout the 1890s. Morgan rejected Weismann's views as placing too much emphasis on phyletic history, as appealing to a "shadowy past" which explained nothing. Certainly heredity plays some role in passing old characters from old cells to new, directed by the chemical substance of the cells. But inheritance cannot explain all. Regeneration shows how the materials of the organism change throughout the whole protoplasm. Regeneration exhibits regulation by way of physiological responses, though the process is so complex that Morgan resisted offering a final theory. Despite his reluctance, he made clear that it is development and response to external conditions rather than internal inherited factors which direct regeneration.

Loeb similarly deemphasized evolutionary or historical factors in discussing fertilization. Artificial fertilization, effected by altered concentrations of salt water, initiates development which can continue to the pluteus stage. These unfertilized eggs have no male component, yet develop normally to that point, demonstrating that whatever the male contributes to heredity is not necessary to initiate development. Loeb thus concentrated on the mechanical and chemical actions, and on the resulting tropisms set up in organisms. In fact, he emphasized that he considered "the chief value of the experiments on artificial parthenogenesis to be the fact that they transfer the problem of fertilization from the realm of morphology into the realm of physical chemistry."[97]

Though these two lectures concerned development, they had each separated out one aspect of development, ignoring general concerns of heredity and evolution. They had specialized. The focus of the other lectures presented had shifted as well. The paper by Daniel Trembly MacDougal

---

97. Jacques Loeb, "On the Nature of the Process of Fertilization," 1899, p. 282.

(1865–1958) on temperature inversion remains an anomaly in the series. It focuses on environment and relatively little on the effects of the climatic conditions on individuals. The other essays on behavior, evolution, and methods all represent different concerns than the *Biological Lectures* had pursued before.

The lectures by Edward Thorndike (1874–1949) reflect Whitman's new central interests at that time as well as the successes of the MBL's neurobiology course. That course, entitled the "Neurological Seminar," was directed by Howard Ayers (1861–1933) during its three-year run, 1896–1899. Designed "for the benefit of investigators who were willing to report the results of unpublished researches on nerve-tissue, and to summarize and discuss the literature bearing on each problem thus reported," the course became what we would today consider as increasingly psychological in content. Many questions of behavior, of mind, of pain, of instinct emerged in that seminar, and several of the papers found their way into the evening lecture series and on into the *Biological Lectures*. Thorndike's discussion of instinct held the general interest in such a multiple role.[98] "Instincts," he wrote, "are the expressions of structures and functions of the nervous system," and they "are as real and as important for the biologist as are bones and blood vessels." In all essential respects, Thorndike seconded Whitman's conclusions. In his second published lecture considering *Paramecium* behavior, he reinforced another of Whitman's pet assumptions by stressing that organisms act as a whole, not as the substance of which they consist. Comparing lower with higher organisms may well bring understanding of seemingly complex processes, as Whitman also assumed.

Another essay on *Paramecium*, by Herbert Spencer Jennings (1868–1947), covered some of the same ground but emphasized the importance of responses to stimuli. Using simple unicellular organisms, the researcher can uncover facts about higher processes as well. Thus, the apparent psychic powers of *Paramecium caudatum* dissolve into expressions of positive chemotaxis, Jennings concluded. Thereby "a long step is taken toward that analysis of vital processes into simple chemical and physical ones, which is deemed by many the final goal of biological science."[99]

98. MBL *Annual Reports* (1899–1903), p. 66.

99. Herbert Spencer Jennings, *Behavior of the Lower Organisms* (Bloomington: Indiana University Press, 1962); original 1906; Philip Pauly, "The Loeb-Jennings Debate and the Science of Animal Behavior," *Journal of the History of the Behavioral Sciences* (1981) 17: 504–515.

Jennings thus agreed with Loeb that biology should seek physico-chemical explanations, but by 1899 the two had begun to disagree about what that meant. Loeb emphasized internal tropisms, set up in the body as gradients of various physical and chemical factors. Jennings saw that some phenomena at least occurred more generally, elicited by various stimuli. Accordingly, he found the stimulus and response less specific. Though only suggested in his 1899 lecture, Jennings' conclusions were developed into a full-scale challenge of Loeb's theories by the time his book appeared in 1906. Though not an MBL regular visitor as most of the other lecturers were, Jennings clearly presented a point of view and offered experimental results directly centered on problems of interest to some of the MBL audience.

Charles Benedict Davenport (1866–1944) considered method in morphology in his lecture. Experimentation is nothing new for morphology and other studies, he pointed out, "but there has been a decided advance upon the methods in vogue a century ago." Notably, application of quantitative methods to zoology is an important addition. Do not fear that the step into the laboratory to count variations and establish frequencies will ruin the charm of biological work, he assured his audience, for one still must go out of doors to gather the specimens for study. Enthusiastic in his optimistic hopes for quantitative methods, Davenport suggested that "in the application of combined experimental and statistical methods to genetic [that is, those concerned with genesis or development] problems, zoology will reach its highest development."[100] Though perhaps not all ears at the MBL remained deaf to Davenport's message, research did not make any significant move in the direction of quantitative study for at least another decade. Population studies or frequency comparisons never characterized much of the work done at the MBL, largely because the MBL emphasized invertebrate and marine work. The populations of interest to most biologists did not come from the sea and were not those roaming individual invertebrates which the MBL researchers found most interesting.

CONCLUSION

The Americans, it should be clear by now, saw a complex of related problems of heredity, individual development, and evolution, or group devel-

---

100. Charles Davenport, "The Aims of the Quantitative Study of Variation," 1899, pp. 267, 270, 272.

opment. They rejected what they regarded as the simplistic theories of Weismann, Roux, and others. Those theories did not fit well with all the evidence at hand, or they failed to explain important facts, the Americans thought. Solid observation, comparison, careful consideration of alternative hypotheses—these features characterized the bulk of the work at the MBL. Perhaps the fact that researchers with different viewpoints and different emphases came together each summer and shared ideas and results served as a corrective to extremism. Over and over we find one or another researcher stressing the necessity of avoiding extreme positions. We find conclusions that what happens is a combination of various factors rather than just one. Especially as the Americans presented public lectures to mixed audiences and worked together teaching courses, they were forced to communicate, to cooperate, to achieve something of that union of specialist laborers that Whitman envisioned.

The 1890s brought hopes for a unified, cooperative biological science. Whitman believed in such a biology which would go beyond morphology and physiology; beyond zoology and botany; beyond embryology, evolution, and heredity. The Marine Biological Laboratory was intended to produce such a cooperative result, to illustrate that biology was one science. In retrospect, the effort did not really succeed. Much of what is properly biological never received attention at the MBL. As the decade moved on, individual investigators began to diverge in their research emphases and increasingly to specialize. Biology never quite became one science. Yet the efforts to address questions of general concern, as revealed in the *Biological Lectures*, and the fact that some questions *were* of general concern, demonstrates that researchers at the MBL in the 1890s at least hoped for a unified biology, even if they recognized that they had not quite achieved it and even if some believed that they might not.

LIST OF BIOLOGICAL LECTURES

1890 (PUBLISHED IN 1890)

1. Charles Otis Whitman, "Specialization and Organization: Companion Principles of All Progress—The Most Important Need of American Biology," 1–26.

2. Charles Otis Whitman, "The Naturalist's Occupation: 1. General Survey. 2. A Special Problem," 27–52.

3. Edmund Beecher Wilson, "Some Problems of Annelid Morphology," 53–78.

4. John Playfair McMurrich, "The Gastraea Theory and Its Successors," 79–106.

5. Edward Gardiner Gardiner, "Weismann and Maupas on the Origin of Death," 107–129.

6. Henry Fairfield Osborn, "Evolution and Heredity," 130–141.

7. Thomas Hunt Morgan, "The Relationships of Sea-Spiders," 142–167.

8. Shosaburo Watase, "On Caryokinesis," 168–187.

9. Howard Ayers, "The Ear of Man: Its Past, Present, and Future," 188–230.

10. William Libbey, "The Study of Ocean Temperatures and Currents," 231–250.

1891 (UNPUBLISHED)

1. Howard Ayers, "The Morphology of the Ear."

2. Henry Herbert Donaldson, "Methods of Studying the Nervous System."

3. J. E. Humphrey, "The Morphology of the Saprolegniaceae."

4. Edward O. Jordan, "Biological Analysis of Water."

5. John Sterling Kingsley, "A Trip to the Bad Lands."

6. Warren Plimpton Lombard, "Some of the Influences Which Affect the Strength of Voluntary Muscular Contraction."
7. John Playfair McMurrich, "The Significance of the Blastopore."
8. Henry Fairfield Osborn, "The Evolution of the Mammalia."
9. Shosaburo Watase, "The Role of the Asters in the Division of the Nucleus."

1892 (UNPUBLISHED)

1. Henry P. Bowditch, "The Determination of Types in Biology."
2. Warren Plimpton Lombard, "Reflex Action" and "Fatigue."
3. Edward Sylvester Morse, "China" and "Japan."
4. William Libbey, "The Physical Geography of the Sea" and "A Trip to Mexico."
5. Henry Fairfield Osborn, "Some Problems of Heredity."
6. Edmund Beecher Wilson, "The Latest Advances in Embryology."
7. William Sedgwick, "Some Problems in Practical Biology."
8. Henry van Peters Wilson, "The Development of Sponges."
9. Howard Ayers, "The Morphology and Physiology of the Ear."
10. Samuel Hubbard Scudder, "The Scales of Butterflies."
11. Henry Herbert Donaldson, "The Architecture of the Brain."
12. Edward O. Jordan, "The History of Bacteriology."
13. Shosaburo Watase, "The Phenomena of Sex Differentiation."

1893 (PUBLISHED IN 1894)

*1. Edmund Beecher Wilson, "The Mosaic Theory of Development," 1–14.
2. Edwin Grant Conklin, "The Fertilization of the Ovum," 15–35.
3. Jacques Loeb, "On Some Facts and Principles of Physiological Morphology," 37–61.
4. John Ryder, "Dynamics of Evolution," 63–81.
5. Shosaburo Watase, "On the Nature of Cell-Organization," 83–103.

* Asterisks designate those lectures that are reprinted in this volume.

6. Charles Otis Whitman, "The Inadequacy of the Cell-Theory of Development," 105-124.

7. Hopkins Marine Laboratory, "Bdellostoma Dombeyi, Lac.," 125-161.

8. W. P. Wilson, "The Influence of External Conditions on Plant Life," 163-183.

9. J. Muirhead Macfarlane, "Irrito-Contractility in Plants," 185-209.

10. Bashford Dean, "The Marine Biological Stations of Europe," 211-234.

11. Charles Otis Whitman, "The Work and the Aims of the Marine Biological Laboratory," 235-242.

1894 (PUBLISHED IN 1895)

1. Amos E. Dolbear, "Life from a Physical Standpoint," 1-21.

2. John Ryder, "A Dynamical Hypothesis of Inheritance," 23-54.

3. Jacques Loeb, "On the Limits of Divisibility of Living Matter," 55-65.

4. George Baur, "The Differentiation of Species on the Galapagos Islands and the Origin of the Group," 67-78.

*5. Henry Fairfield Osborn, "The Hereditary Mechanism and the Search for the Unknown Factors of Evolution," 79-100.

6. Edmund Beecher Wilson, "The Embryological Criterion of Homology," 101-124.

7. John Playfair McMurrich, "Cell-Division and Development," 125-147.

*8. Wilhelm Roux, translated by William Morton Wheeler, "The Problems, Methods, and Scope of Developmental Mechanics," 149-190.

9. J. M. Macfarlane, "The Organization of Botanical Museums for Schools, Colleges, and Universities," 191-204.

10. Charles Otis Whitman, "Evolution and Epigenesis," 205-224.

11. Charles Otis Whitman, "Bonnet's Theory of Evolution," 225-240.

12. Charles Otis Whitman, "The Palingenesis and the Germ Doctrine of Bonnet," 241-272.

13. Shosaburo Watase, "Origin of the Centrosome," 273-287.

1895 (PUBLISHED IN 1896)

1. Simon Flexner, "Infection and Intoxification," 1-10.

2. George Sternberg, "Immunity," 11-28.

3. Henry Fairfield Osborn, "A Student's Reminiscence of Huxley," 29–42.

4. William Berryman Scott, "Paleontology as a Morphological Discipline," 43–61.

5. Amos E. Dolbear, "Explanations, or How Phenomena Are Interpreted," 63–82.

6. Amos E. Dolbear, "Known Relations between Mind and Matter," 83–99.

7. Shosaburo Watase, "On the Physical Basis of Animal Phosphorescence," 101–118.

8. William Locy, "The Primary Segmentation of the Vertebrate Head," 119–136.

9. J. S. Kingsley, "The Segmentation of the Head," 137–148.

10. Charles Sedgwick Minot, "Bibliography—A Study of Resources," 149–168.

11. George Atkinson, "The Transformation of Sporophylla to Vegetative Organs," 169–188.

1896–1897 (PUBLISHED IN 1898)

1. Hermann C. Bumpus, "The Variations and Mutations of the Introduced Sparrow, *Passer Domesticus*," 1–15.

*2. Edwin Grant Conklin, "Cleavage and Differentiation," 17–43.

3. Katherine Foot, "The Centrosomes of the Fertilized Egg of Allolobophora Foetida," 45–57.

4. William Berryman Scott, "The Methods of Paleontological Inquiry," 59–78.

5. Arnold Graf, "The Physiology of Excretion," 79–107.

6. Burt Wilder, "Some Neural Terms," 109–173.

7. D. P. Penhallow, "A Classification of the North American Taxaceae and Coniferae on the Basis of the Stem Structure," 175–192.

8. James Ellis Humphrey, "The Selection of Plant Types for the General Biology Course," 193–202.

9. Albert Davis Mead, "The Rate of Cell-Division and the Function of the Centrosome," 203–218.

10. Henry E. Crampton, "Coalescence Experiments upon the Lepidoptera," 219–229.

11. Charles Otis Whitman, "Some of the Functions and Features of a Biological Station," 231–242.

1898 (PUBLISHED IN 1899)

1. Edmund Beecher Wilson, "The Structure of Protoplasm," 1–20.

2. Edmund Beecher Wilson, "Cell-Lineage and Ancestral Reminiscence," 21–42.

3. Frank Rattray Lillie, "Adaptation in Cleavage," pp. 43–67.

4. Edwin Grant Conklin, "Protoplasmic Movement as a Factor of Differentiation," 69–92.

5. Aaron Louis Treadwell, "Equal and Unequal Cleavage in Annelids," 93–111.

6. Albert Davis Mead, "The Cell Origin of the Prototroch," 113–138.

*7. Cornelia Maria Clapp, "Relation of the Axis of the Embryo to the First Cleavage Plane," 139–151.

8. Thomas H. Montgomery, "Observations on Various Nucleolar Structures of the Cell," 153–175.

9. Shosaburo Watase, "Protoplasmic Contractility and Phosphorescence," 177–192.

10. Thomas Hunt Morgan, "Some Problems of Regeneration," 193–207.

11. Herman C. Bumpus, "The Elimination of the Unfit as Ilustrated by the Introduced Sparrow, *Passer Domesticus*," 209–226.

12. Jacques Loeb, "On the Heredity of the Marking in Fish Embryos," 227–234.

13. W. W. Norman, "Do the Reactions of Lower Animals due to Injury Indicate Pain-Sensations?," 235–241.

14. William Berryman Scott, "North American Ruminant-like Mammals," 243–264.

*15. William Morton Wheeler, "Caspar Friedrich Wolff and the *Theoria Generationis*," 265–284.

*16. Charles Otis Whitman, "Animal Behavior," 285–338.

1899 (PUBLISHED IN 1900)

1. Douglas P. Campbell, "The Evolution of the Sporophyte in the Higher Plants," 1–18.

 2. D. P. Penhallow, "The Nature of the Evidence Exhibited by Fossil Plants, and Its Bearing upon Our Knowledge of the History of Plant Life," 19-35.

 3. Daniel T. MacDougal, "Influence of Inversions of Temperature, Ascending and Descending Currents of Air, upon Distribution," 37-47.

 4. Daniel T. MacDougal, "Significance of Mycorrhizas," 49-56.

 5. Edward Thorndike, "Instinct," 57-67.

 6. Edward Thorndike, "The Associated Processes in Animals," 69-91.

 *7. Herbert Spencer Jennings, "The Behavior of Unicellular Organisms," 93-112.

 8. Carl Eigenmann, "The Blind-fishes," 113-126.

 9. Alpheus Hyatt, "Some Governing Factors Usually Neglected in Biological Investigations," 127-156.

10. A. G. Mayer, "On the Development of Color in Moths and Butterflies," 157-160.

11. Albert Prescott Mathews, "The Physiology of Secretion," 165-183.

*12. Thomas Hunt Morgan, "Regeneration: Old and New Interpretations," 185-208.

13. Gary Calkins, "Nuclear Division in Protozoa," 209-230.

14. Charles Manning Child, "The Significance of the Spiral Type of Cleavage and Its Relation to the Process of Differentiation," 231-266.

15. Charles B. Davenport, "The Aims of the Quantitative Study of Variation," 267-272.

*16. Jacques Loeb, "On the Nature of the Process of Fertilization," 273-282.

Opposite: The first MBL buildings, illustrating the growth from one building in 1888 to two joined sections in 1890 to three sections in 1892. Every year Whitman argued with the Board of Trustees over whether they would add new facilities, and every two years for the first decade the trustees reluctantly added another building. This first unit was affectionately called "Old Main" and remained standing until 1970, when progress dictated its replacement with a more modern facility.

The motto from Louis Agassiz's Anderson School on Penikese Island, handed down to the MBL and placed here above the library, which was housed in the corner of the first building. The plaque now hangs above the entrance to the library stacks in the Lillie Building.

The interior of the laboratory during the first session, 1888. From front to back on the right: Edward Gardiner Gardiner, investigator, MIT; Mrs. Helen Torrey Harris, investigator, Wellesley College; William Thompson Sedgewick, visiting lecturer, MIT; Charles Otis Whitman, director; Marcella I. O'Grady (later Mrs. Theodore Boveri), investigator, Bryn Mawr; Cornelia Clapp, investigator, Mt. Holyoke; Mrs. Susan J. Hart, student, Jackson, Michigan, high school teacher; Edwin O. Jordan, student, MIT; Isabel Mulford, investigator, Vassar. On the left: James H. Norton, student, Ravenswood, Illinois, high school teacher; F. L. Washburn, investigator, University of Michigan.

The MBL group had fun as well as work, including a visit in 1890 from the touring dancing bear Josephine, here in front of Old Main.

A typical group shot, of which the MBL has many in its archival collection. This one shows, left to right, front row: William Morton Wheeler, William Setchell, Charles Otis Whitman, Hermon C. Bumpus, and Shosaburo Watase. Back row: Pierre Fish, Jacques Loeb, Edwin O. Jordan, Charles Bristol, and Edwin Grant Conklin.

View across the Eel Pond, in front of Old Main, 1894. Today this spot is occupied by parking lots and boat docks, and the pond is filled with boats year-round.

Returning from the annual MBL picnic, where everyone gathered and chugged off to one of the nearby islands for a day of relaxing, collecting, and exploring. Such events helped to create the family feeling that drew so many of the MBL participants to Woods Hole summer after summer.

The group portrait in 1895 showing the "Vigilant," on which the group relied for collecting marine materials.

Ctenophore collecting at nearby Quisett Harbor, 1897. The woman in front and the man behind her holding up the collecting jar are Gertrude Stein and her brother Leo.

Clearly the number of people who carried out their "spasmodic descent upon the seashore," as E. Ray Lankester put it, had increased considerably beyond the original seventeen by the time this group photograph was taken in 1895.

THE LECTURES

I

EDMUND BEECHER WILSON
1856–1939

Wilson received a Ph.B. degree from Yale University's Sheffield Scientific School and a Ph.D. from Johns Hopkins. His introduction to marine study began in 1877 with an appointment to the United States Fish Commission under Spencer Fullerton Baird and continued with his work at Hopkins' Chesapeake Zoological Station, run by William Keith Brooks. In 1882–1883 he began study at Cambridge, then worked at Rudolf Leuckart's laboratory in Leipzig and at the Naples Zoological Station. His return to the United States took him to positions at Williams College, Massachusetts Institute of Technology, Bryn Mawr, and after 1891 to Columbia University, where he stayed. There, his work in cytology gained international attention so that Wilson was soon regarded as one of America's leading biologists. His textbook, *The Cell in Development and Inheritance*, lent support to his growing international reputation.

Wilson first attended the MBL session in 1889, then continued to spend virtually every summer there. By 1890 he had been elected a Trustee, and thereafter he always played an active role in MBL affairs. He also presented several lectures, the quality of which the following exemplifies. The first lectures appeared as more traditionally morphological, concerned with geneological relationships among organisms. This one represents Wilson's move to consider thoroughly modern topics, especially the repeatedly discussed mosaic theory of development and experimentation in biology. Yet he also retained a careful approach, not rejecting tradition but interpreting and modifying it. Wilson's attitudes, while more clearly articulated than those of some of his contemporaries, reflect the general excitement about embryology with the simultaneous rejection of the particular German mosaic theory. That theory Wilson and most others at the MBL regarded as moving too far beyond the desired "*terra firma* of observed fact."

EDMUND BEECHER WILSON.

THE MOSAIC THEORY OF DEVELOPMENT

EDMUND B. WILSON

A REMARKABLE awakening of interest and change of opinion
has of late taken place among working embryologists in regard
to the cleavage of the ovum.   So long as the study of embryology was dominated by the so-called biogenetic law, so long
as the main motive of investigation was the search for phyletic
relationships and the construction of systems of classification,
the earlier stages of development were little heeded.   The
two-layered gastrula was for the most part taken as the real
starting-point for research, and the segmentation stages were
briefly dismissed as having little purport for the more serious
problems involved in the investigation of later stages.   The
cleavage is equal or unequal, total or partial, regular or irregular;
the diblastic condition attained by delamination, migration or
invagination ; the gastrulation embolic or epibolic : — such
were the general conclusions announced regarding the præ-
gastrular stages in a large proportion of the embryological
papers published down to the time of Balfour and even later.
The last decade has, however, witnessed so extraordinary a
change of front on this subject that it will not be out of place
to review briefly the three leading causes by which it has been
brought to pass.

First, it has become more and more clear that the germ-layer
theory is, to a certain extent, inadequate and misleading, and
that even the primary layers of the "gastrula" cannot be
regarded as strictly homologous throughout the animal kingdom.
To assume that they are so involves us in inextricable difficulties — such as those for instance encountered in the comparison
of the annelid gastrula with that of the chordates, or the com-

parison of the sexual and asexual modes of development in tunicates, bryozoa, worms and cœlenterates. This considera- tion led some morphologists to insist on the need of a more precise investigation of the præ-gastrular stages, and the desi- rability of taking as a starting-point not the two-layered gastrula but the undivided ovum. " The ' gastrula ' cannot be taken as a starting-point for the investigation of comparative organo- geny unless we are certain that the two layers are everywhere homologous. Simply to assume this homology is simply to beg the question. *The relationship of the inner and outer layers in the various forms of gastrulas must be investigated not only by determining their relationship to the adult body, but also by tracing out the cell-lineage or cytogeny of the individual blastomeres from the beginning of development.*"

The second of the causes referred to was the discovery of the so-called pro-morphological relations of the segmenting ovum. It is now just ten years since Roux and Pflüger independently announced the discovery that the first plane of cleavage in the frog's egg coincides with the median plane of the adult body (a fact announced many years earlier by Newport, whose obser- vation fell, however, into oblivion). The same result was soon afterwards reached in the case of the cephalopod (Watase) and tunicate (Van Benden and Julin), and for a time it seemed not improbable that a general law had been determined. Later researches disappointed this expectation ; for it was demon- strated that the first cleavage plane may be transverse to the body (annelids, gasteropods, urodeles), or even in some cases show a purely variable and inconstant relation (teleosts). The fact remained, however, that in the greater number of known cases definite relations of symmetry can be made out between the early cleavage stages and the adult body ; and this fact invested these stages with a new and captivating interest.

The third and most important cause lay in the new and startling results attained by the application of experimental methods to embryological study, and especially to the investi- gation of cleavage. The initial impulse in this direction was given in 1883 by the investigations of Pflüger upon the influence of gravity and mechanical pressure upon the segmenting ova of the frog.   These pioneer studies formed the starting-point for a series of remarkable researches by Roux, Driesch, Born and others, that have absorbed a large share of interest on the part of morphologists and physiologists alike ; and it is perhaps not too much to say that at the present day the questions raised by these experimental researches on cleavage stand foremost in the arena of biological discussion, and have for the time being thrown into the background many problems which were but yesterday generally regarded as the burning questions of the time.   It is the purpose of this lecture to consider, briefly, the most central and fundamental subject of the current controversy.

It is an interesting illustration of how even scientific history repeats itself that the leading issue of to-day has many points of similarity to that raised two hundred years ago between the præ-formationists and the epigenesists.   Many leading biological thinkers now find themselves compelled to accept a view that has somewhat in common with the theory of præ-formation, though differing radically from its early form as held by Bonnet and other evolutionists of the eighteenth century.   No one would now maintain the archaic view that the embryo præ-exists *as such* in the ovum.   Every one of its hereditary characters is, however, believed to be represented by definite structural units in the idioplasm of the germ-cell, which is therefore conceived as a kind of microcosm, not similar to, but a perfect symbol of, the macrocosm to which it gives rise (Hertwig).   In its modern form this doctrine was first clearly set forth by Darwin in the theory of Pangenésis ('68).   Twenty years later ('89) it was remodeled and given new life by Hugo de Vries, in a profoundly interesting treatise entitled *Intracellular Pangenesis* and in its new form was accepted by Oscar Hertwig, and pushed to its uttermost logical limit by Weismann.   Kindred theories have been maintained by many other leading naturalists.

The considerations which have led to the rehabilitation of the theory of pangenesis are based upon the facts of what

Galton has called *particulate inheritance.* The phenomena of atavism, the characters of hybrids, the facts of spontaneous variation, all show that even the most minute characteristics may independently appear or disappear, may independently vary, and may independently be inherited from either parent without in any way disturbing the equilibrium of the organism, or showing any correlation with other variations. These facts, it is argued, compel the belief that hereditary characteristics are represented in the idioplasm by distinct and definite germs ("pangens," "idioblasts," "biophores," *etc.*), which may vary, appear or disappear, become active or latent, without affecting the general architecture of the substance of which they form a part. Under any other theory we must suppose variations to be caused by changes in the molecular composition of the idioplasm as a whole, and no writer has shown, even in the most approximate manner, how particulate inheritance can thus be conceived.

Based upon this conception two radically different theories of development have recently been propounded. The first of these — the so-called mosaic theory of Roux and Weismann, which forms the subject of this lecture — is based upon the assumption that the cause of differentiation lies in the nature of cell-division. Karyokinesis is conceived as qualitative in character in such wise that the idioplasmic germs are sifted apart, and cells of different prospective values receive their appropriate specific germs at the moment of their formation. The idioplasm therefore becomes progressively simpler as the ontogeny goes forward, except in the case of the germ-cells; these retain a store of the original mixture ("germ-plasm" of Weismann). Every cell must therefore possess an independent power of self-determination inherent in the specific structure of its idioplasm, and the entire ontogeny is aptly compared by Roux to a mosaic-work; it is essentially a whole arising from a number of independent self-determining parts, though Roux qualifies this conception by the admission that the self-determining power of the cell is capable in some measure, of modification, through interaction with its fellows ("correlative differentiation").

In the hands of Weismann this theory attains truly colossal proportions. The primary germs or units (which he calls "biophores") are aggregated to form "determinants," the determinants to form "ids," and the ids to form "idants," which are identified with the chromosomes of the ordinary karyokinetic figure. Upon this basis is reared a stately group of theories relating to reproduction, variation, inheritance and regeneration, which are boldly pushed to their utmost logical limit. These theories await the judgment of the future. Brilliantly elaborated and persuasively presented as they are, they do not at present, I believe, carry conviction to the minds of most naturalists, but arouse a feeling of scepticism and uncertainty; for the fine-spun thread of theory leads us little by little into an unknown region, so remote from the *terra firma* of observed fact that verification and disproof are alike impossible.

In its original form the mosaic theory has, I believe, received its death-blow from the facts of experimental embryology, though both Roux and Weismann still endeavor to maintain their position. It is rather curious that the very line of research struck out by Roux, by which he was led to the mosaic theory, should in later years have ended in a view diametrically opposed to his own. In 1888 Roux succeeded in killing (by puncture with a heated needle) one of the first two blastomeres of the segmenting frog's egg. The uninjured blastomere continued its development as if still forming a part of an entire embryo, giving rise successively to a half-blastula, half-gastrula, and half-tadpole embryo, with a single medullary fold. Analogous results were reached by operation upon four-celled stages. It was this result that led Roux to compare the development to a mosaic-work, asserting that "the development of the frog-gastrula, and of the embryo immediately derived from it is, from the second cleavage onward, a mosaic-work, consisting of at least four vertical independently developing pieces." Roux himself, however, showed that in later stages the missing half (or fourth) is perfectly restored by a process of "post-generation," which begins about the time of the formation of the medullary folds — a result which, in itself, really contradicts the mosaic hypothesis ; for the course of events in the uninjured blastomere, or its products, is radically altered by changes on the other side of the embryo.

A more decisive result was reached in 1891 by Driesch, who succeeded, in the case of *Echinus*, in effecting a complete separation of the blastomeres by shaking them apart. A blastomere of the 2-celled stage, thus isolated, gave rise to a perfect but half-sized blastula, gastrula, and Pluteus larva ; an isolated blastomere of the 4-celled stage produced a perfect dwarf gastrula one-fourth the normal size. Even in this case, however, the earliest stages of development (cleavage) showed traces of the normal development, the isolated blastomere segmenting, as if it were a half-embryo, and only becoming a perfect whole in the blastula stage. In the following year, however, the writer repeated Driesch's experiments in the case of *Amphioxus* (the egg of which is extremely favorable for experiment), and found that in this case there is, as a rule, no preliminary half-development whatever. The isolated blastomere behaves from the beginning like an entire ovum of one-half or one-fourth the normal size.

It is quite clear that in *Amphioxus* the first two divisions of the ovum are not qualitative, as the mosaic theory assumes, but purely quantitative ; for the fact that each of the two or four blastomeres may give rise to a perfect gastrula proves that all contain the same materials. Nevertheless, in the normal development, these cells give rise to different structures — *i. e.*, they have a different prospective value — from which it follows that, in this case at least, differentiation is not caused by qualitative cell-division, but by the conditions under which the cell develops.

These facts are obviously a serious blow to the mosaic theory, and the efforts of Roux and Weismann to sustain their hypothesis in the face of such evidence only serve to emphasize the weakness of their case. In order to explain the facts of post-generation — *i. e.*, the capability of isolated blastomeres to produce complete embryos — both Roux and Weismann are compelled to set up a subsidiary hypothesis, assuming that during cell-division each cell may receive, in addition to its specific form of idioplasm, a portion of unmodified idioplasm afforded by purely quantitative division. This unmodified idioplasm ("accessory idioplasm" of Weismann, or in some cases "germ-plasm"; "post-generation or regeneration idioplasm" of Roux) remains latent in normal development which is controlled by the active specific idioplasm. Injury to the ovum — e. g., mechanical separation of the blastomeres — acts as a stimulus to the latent idioplasm, which thereupon becomes active, and causes a repetition of the original development. By assuming a variable latent period following the stimulus, Roux is able to explain the fact that regeneration takes place at different periods in different animals.

Considered as a purely formal explanation this subsidiary hypothesis is perfectly logical and complete. A little reflection will show, however, that it really abandons the entire mosaic position, by rendering the assumption of qualitative division superfluous; and, aside from this, its forced and artificial character, places a strain upon the mosaic theory under which it breaks down. Both of the two fundamental postulates of the modified theory— viz., qualitative nuclear division, and accessory latent idioplasm — are purely imaginary. They are complicated assumptions in regard to phenomena of which we are really quite ignorant, and they lie at present beyond the reach of investigation. The "explanation" is, therefore, unreal; it carries no conviction, and no real explanation will be possible until we possess more certain knowledge regarding the seat of the idioplasm (which is entirely an open question), and its internal composition and mode of action (which is wholly unknown). In the meantime we certainly are not bound to accept an artificial explanation like that of Roux, however logical and complete, unless it can be shown that the phenomena are not conceivable in any other way.

We turn now to a brief consideration of opposing views, among which I ask attention especially to those of Driesch and Hertwig. In common with Kölliker and many other eminent authorities, these authors insist that cell-division is not qualitative but quantitative only, and hence is not, *per se*, a cause of differentiation, for there is no sifting apart of the idioplasmic units, but an equal distribution of them to all the cells of the body. In other words, the cleavage of the ovum does not effect an analysis of the idioplasm into its constituent elements, but only breaks it up into a large number of similar masses. Differentiation follows upon cell-division, is caused by the interaction of the parts of the embryo, and the character of the individual cell is determined by its environment — *i. e.*, by its relation to the whole of which it forms a part. "The egg," says Hertwig, "is an organism, which multiplies by division to form numerous organisms equivalent to itself, and it is through the interactions of all these elementary organisms, at every stage of the development, that the embryo, as a whole, undergoes progressive differentiation. The development of a living creature is therefore in no wise a mosaic work, but, on the contrary, all the individual parts develop in constant relation one to another, and the development of the part is always dependent on the development of the whole." There is therefore no necessary relation between the individual blastomeres of the segmenting ovum and the parts of the adult body to which they give rise ; this relation is purely fortuitous. The most extreme statement of this view appears in the writings of Pflüger and Driesch. "I would accordingly conceive," says Pflüger, "that the fertilized egg has no more essential relation to the later organization of the animal than the snowflake has to the size and form of the avalanche which, under appropriate conditions, may develop out of it." Driesch, writing ten years later ('89), is no less explicit. He regards the blastomeres of the *Echinus* embryo, as "composed of an indifferent material, so that they may be thrown about at will, like balls in a pile, without the least impairment of their power of development." The ultimate fate of any particular blastomere is determined by its relative position in the mass ; that is (to quote his own striking aphorism), "their prospective value (*Bedeutung*) is a function of their location" (cf. His).

We shall presently return to these more extreme views, but I will here point out one all-important point which is definitely established by the work of Driesch and other experimentalists, and which is accepted by all opponents of the mosaic theory, namely, that the cell cannot be regarded as an isolated and independent unit.   The only real unity is that of the entire organism, and as long as its cells remain in continuity they are to be regarded, not as morphological individuals, but as specialized centres of action into which the living body resolves itself, and by means of which the physiological division of labor is effected.   This view, at which a number of embryologists have independently arrived, has been most ably urged by Whitman, in one of the lectures of this volume, though in connection with a general conception of development peculiarly his own.

It is important not to lose sight of the fact that Hertwig, no less than Roux and Weismann, conceives the idioplasm (which he would locate in the cell-nucleus) as an aggregate of units ("idioblasts") which severally correspond to the hereditary qualities of the organism ; and since cell-division is not qualitative, every cell must contain the sum total of the hereditary character of the species.   Differentiation is conceived by Hertwig (following de Vries) as the result of physiological changes in the idioblasts, some of which remain latent, while others become active, and thus determine the specific character of the cell, according to the nature of the active idioblasts.   In regeneration such of the latent idioblasts are called into action as are necessary to carry out the regenerative process.

We have found good reason for the conclusion that the mosaic theory cannot, in its extreme form, be maintained.   It remains to inquire whether the extreme anti-mosaic conception rests upon a more secure foundation, and whether the mosaic hypothesis may not contain certain elements of truth.   I have elsewhere more than once pointed out that the views of Hertwig and Driesch have received a strong bias, from the circumstance that the discussion has hitherto been confined mainly to the echinoderm egg, which shows no visible differentiation in the cells until a relatively late period (16-celled stage).

The whole question assumes a somewhat different aspect when we regard such highly differentiated types of cleavage as we find, for example, among the annelids ; and I would ask attention for a moment to the case of *Nereis*, which is, at present, the best known form. Differentiation here begins at the very first cleavage (which is conspicuously unequal), and it becomes more pronounced with every succeeding division. The median plane is marked out at the second cleavage ; at the third the entire ectoblast of the trochal and præ-trochal regions is formed ; at the fourth the material for the entire "ventral plate" (including the ventral nerve-cord and the seta-sacs) is segregated in a single cell, that for the stomodæum in three cells ; the fifth cleavage completes the ectoblast, and by the 38-celled stage the germ-layers are completely segregated (the mesoblast in a single cell) and the architecture of the embryo is fully outlined in the arrangement of the parent blastomeres, or protoblasts.

We do not know whether, in this case, the first two blasto-meres are qualitatively different, though there may be some ground for holding that they are, from the fact that the larger of the two contains a relatively larger proportion of protoplasm than the smaller.[1] But in any case their difference in size renders it impossible that they should play interchangeable parts in the cleavage. The entire later development is, how-ever, moulded upon the 2-celled stage, every blastomere having a definite relation to it and a definite morphological value. The development is a visible mosaic-work, not one ideally conceived by a mental projection of the adult characteristics back upon the cleavage stages. The principle of "organbildende Keim-bezirke" has here a real meaning and value, and this would remain true even if it should hereafter be shown that both of the first two blastomeres of *Nereis*, if isolated, could produce a perfect embryo.

It is clear, from such a case, that the more extreme views of Driesch and Hertwig cannot be accepted without consider-able modification. It seems to me, however, that they may be modified in such a way as, without sacrificing the principle of epigenesis for which they contend, to recognize certain ele-

[1] All my attempts to separate these blastomeres by shaking have thus far been unsuccessful.

ments of truth in the mosaic hypothesis ; and I will attempt to indicate this modification by a comparison between *Amphioxus* and *Nereis*. In the case of *Amphioxus* we have the clearest evidence that differentiation is, in a measure, dependent upon the relation of the cell to the whole of which it forms a part. The first visible differentiation in this case is at the third cleavage, which consists in an unequal division of each of the four blastomeres, so as to give rise to four micromeres and four macromeres, the former giving rise to ectoblast only, while the latter give rise to entoblast and mesoblast as well (*Diagram* I). If, however, the blastomeres of the 4-celled stage be separated (shaken apart) the course of events is entirely changed ;. for in this case each divides equally, not unequally, and ultimately gives rise to a complete quarter-sized dwarf, instead of one-quarter of a

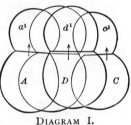

DIAGRAM I.

normal embryo, as it would have done under ordinary circumstances. The character of the fourth cleavage is here directly or indirectly determined in each cell by the relation of the cell to its fellows ; and if this is true of any one stage of the ontogeny, a very strong presumption is created that it is true of all — that, in the process of progressive differentiation occurring in the course of every animal ontogeny, the character of each step is determined by the condition of the entire organism. The ontogeny is, in other words, a connected series of interactions between the various parts of the embryo, in which each step establishes new relations, through which the following step is determined. The character of the series, as a whole, depends upon the first step, and this in turn upon the constitution of the original ovum. In *Amphioxus* differentiation proceeds slowly, the earlier blastomeres show no appreciable divergence, and the first stages show no trace of a mosaic work. In *Nereis*, on the other hand, a mosaic-like character appears from the beginning, because of the inequality of the first cleavage, which conditions the entire subsequent development through the peculiar relations established by it.   The cause of the inequality must lie in the undivided ovum, and a study of the first cleavage-spindle shows that the inequality is unmistakably foreshadowed before the least outward sign of division appears ; for the asters at the spindle-poles are conspicuously unequal in size, the larger aster corresponding with the future larger cell (Diagram II).   This difference is not connected with any

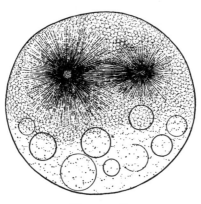

determinable mechanical conditions ; for the centrosomes lie nearly equidistant from the membrane (the egg is spherical), and the deutoplasm shows no perceptible inequality in horizontal distribution.   The conclusion seems unavoidable that the differentiation in size is caused by a specific form of activity in the cytoplasm (or archoplasm),

DIAGRAM II.

occurring prior to cell-division.   But if a differentiation in size may have such an origin, we may fairly argue that other differentiations may likewise precede cell-division, and that in such cases the division may be, in a sense, qualitative.

It seems to me, that in these considerations we may find, in some measure, a reconciliation between the extremes of both the rival theories under discussion — that we may consistently hold with Driesch that the prospective value of a cell may be a function of its location, and at the same time hold with Roux that the cell has, in some measure, an independent power of self-determination due to its inherent specific structure.   Such a view is only possible, however, if we regard the specific structure of the cell to have arisen not through the segregation and isolation within its boundaries of special idioblasts or germsubstances, that have been sifted out by qualitative division, but through a physiological specialization (as de Vries and Hertwig insist) that may have taken place before, during, or after cell-division, according to circumstances.   If differentiation precedes or accompanies division, the latter process may be in a sense qualitative. If it follows, division will be purely quantitative, and in such a case we may rightly speak of differentiation as a result of cellular interaction. The segmentation of the egg presents more or less of a mosaic-like character, according to the period at which differentiation appears, and the rate at which it proceeds, as expressed in limitations of the power of development in the individual blastomeres, and their differences in size and structure.

The general interpretation of development which I have thus endeavored to sketch will be found to differ widely in some respects from that set forth in one of the subsequent lectures of this volume, from which, through Professor Whitman's courtesy, I am enabled to quote. Whitman argues that "cell-orientation may enable us to infer organization, but to regard it as a measure of organization is a serious error." "The question as to the presence of organization," he says, "is not settled by the *form* of cleavage. Eggs that admit of complete orientation at the first or second cleavage, or even before cleavage begins, are commonly supposed to reflect *precociously* the later organization, while eggs in which such early orientation is impossible are supposed to be more or less completely isotropic and destitute of organization. When the region of apical growth is represented by conspicuous teloblasts, the fate of which is seen to be definitely fixed from the moment of their appearance, we find it impossible to doubt the evidence of organization, or 'precocious differentiation' as it is conventionally called. When the same region is composed of more numerous cells, among which we are unable to distinguish special proliferating cells, we lapse into the irrational conviction that the absence of definitely orientable cells means just so much less organization."

It would be manifestly out of place to enter here upon any of the interesting discussions suggested by the passage just quoted, and I will therefore only add that Professor Whitman's position seems to me to rest upon a special and peculiar use of the word "organization," and that his view leads to a denial of the principle of epigenesis. No one would maintain that the living egg is "destitute of organization," but neither can any one maintain that the egg-organization is identical with that of the adult. Development is essentially a transformation of one form of organization into another along the path of cell-division and cell-differentiation ; and it is undeniable that the adult form of organization is thus expressed earlier in some cases than in others — for example, in the segregation of the germ-layers in the polyclade, as compared with the annelid or gasteropod. We are still profoundly ignorant of the nature and causes of differentiation, and of its precise relation to cell-formation ; and the question is probably not yet ripe for discussion. It is, however, impossible to maintain that differentiation in the Metazoa is entirely independent of cell-formation, when we recall the multitude of cases in which the lines of differentiation coincide with cell-boundaries.

2

HENRY FAIRFIELD OBSORN
1857–1935

Often remembered for his later political excesses at the American Museum of Natural History in New York, Osborn considered himself a biologist in his early years. He attended Princeton University (then the College of New Jersey) and there became a lifelong friend of William Berryman Scott, another MBL lecturer. Expeditions to collect fossils in the western states were followed by Osborn's study under Thomas Henry Huxley and Francis Balfour in London. He taught at Princeton, then went to Columbia to develop the department of biology and a department of mammalian paleontology at the American Museum. While at Columbia, Osborn visited the MBL often and served on the Board of Trustees. Yet correspondence reveals that he never stayed at the MBL long, certainly not as long as Whitman would have liked. Whitman cared centrally about evolution, and felt keenly the importance of maintaining an evolutionary perspective on biological questions. Thus, Whitman urged the study of life history and behavior of organisms as well as embryological and physiological study. Osborn brought to the MBL that important historical outlook on heredity and development.

Indeed, Osborn regarded heredity, development, and evolution as closely connected phenomena. In this lecture of 1894, he reviewed the current theories of heredity by Herbert Spencer and August Weismann and rejected both as speculative excesses. In his survey and assessment of available alternative theories, he concluded that no one had yet achieved a unified theory of heredity and evolution, a theory which he clearly held as a desirable goal. Experimental evolution might provide a crucial test of the effect of environment on development, Osborn thought, and thus help decide among the theories. This essay, like his others at the MBL, focused attention on the relations among specialists in diverging areas of biological study and encouraged discussion of their varied points of view.

HENRY FAIRFIELD OSBORN.

THE HEREDITARY MECHANISM AND THE
SEARCH FOR THE UNKNOWN FACTORS
OF EVOLUTION [1]

HENRY FAIRFIELD OSBORN

"Disprove Lamarck's principle and we must assume that there is some third factor in Evolution of which we are now ignorant." [2]

Chief among the unknown factors of evolution are the relations which subsist between the various stages of development and the environment.

A STUDY of the recent discussion in the *Contemporary Review* between Spencer and Weismann leads to the conclusion that neither of these acknowledged leaders of biological thought supports his position upon inductive evidence. Each displays his main force in destructive criticism of his opponent; neither presents his case constructively in such a manner as to carry conviction either to his opponent or to others. In short, beneath the surface of fine controversial style we discern these leaders respectively maintaining as finally established, theories which are less grounded upon fact than upon the logical improbabilities of rival theories. Such a conclusion is deeply significant; to my mind it marks a turning point in the history of speculation, for certainly we shall not arrest research with any evolution factor grounded upon logic rather than upon inductive demonstration. A retrograde chapter in the history

---

[1] This lecture is mainly from an article published by the author, in Merkel u. Bonnet: *Ergebnisse für Anatomie und Entwickelungsgeschichte*, Freiburg, 1894, and partly from a paper before the Biological Section of the British Association for the Advancement of Science: Certain Principles of Progressively Adaptive Variation observed in Fossil Series. *Nature*, August 30, 1894.

[2] Osborn: Are Acquired Variations Inherited? Address before the American Society of Naturalists. *Amer. Naturalist*, February, 1891.

of science would open if we should do so and should accept as established laws which rest so largely upon negative reasoning.

The growing sentiment of the necessity of induction and of inductive evidence is the least conspicuous, but really the most important and lasting outcome of this prolonged discussion. Weismann is the real initiator of this outcoming movement although it has taken a radical direction he neither foresaw nor advocated, for his position is eminently conservative. In fact his first permanent service to Biology is his demand for direct evidence of the Lamarckian principle, which has led to the counter-demand for such evidence of his own Selection principle, which by his own showing, and still more by his own admission in this discussion with Spencer, he is unable to meet. His second permanent service, as Professor Wilson reminds the writer, is that he has brought into the foreground the relation between the hereditary mechanism and evolution.

What have we gained in the controversy of the past decade unless it is closer thinking and this keener appreciation of the necessity for more observation? We carry forth, perhaps, some new and useful working hypotheses as to possible modes of evolution, and a fuller realization of the immense difficulties of the heredity problem — but these are only indirect gains. It is a direct gain that these negative results have led a minority of biologists into a total reaction from speculation and into a generally agnostic temper towards modern theories which is far more healthy and hopeful than the confident spirit of the majority upon either the Neo-Lamarckian or the Neo-Darwinian side. There is no note of progress in the dogmatic assertion that the question is established either as Spencer or as Weismann would have it, unless this assertion can be backed up by proof, and by whom can proof be presented if not by these masters of the subject? The conviction we all reach when we sift wheat from chaff, and bring together from all sources phenomena of different kinds and seek to discern what the exact bearings of these phenomena are, is that we are still on the threshold of the evolution problem, and that the secret is largely tied up with that of vital phenomena in general.

The very wide and positive differences of opinion which pre-

vail are attributable largely to the unnatural divorce of the
different branches of biology, to our extreme modern specialization, to our lack of eclecticism in biology.   We begin to
grasp the magnitude of the problem only when side by side
with field and laboratory data are placed paleontological data,
as well as anthropological, including the unique facts of human
variation and the laws of human inheritance.   For in modern
embryology certainly the most brilliant discovery is that the
physical basis of all inheritance is the same — and growing out
of this is the high probability that the laws of heredity are
the same in the whole organic world, with no barriers between
protozoa and metazoa, or between animals and plants.   Both
Weismann and Spencer show themselves blind to this nexus
of fundamental uniformity when they draw certain lines of
division in inheritance where none exist in the visible hereditary mechanism of chromatin and archoplasm.   With these
discoveries in mind does not Weismann appear as much afield
when he maintains that the inheritance of acquired characters
is a declining principle in the ascent of life, as Spencer when
he maintains that it is a rising principle in the ascent of life?

The first step then towards progress is the straightforward
confession of the limits of our knowledge and of our present
failure to base either Lamarckism or Neo-Darwinism as universal principles upon induction.   The second is the recognition that all our thinking still centers around the five working
hypotheses which have thus far been proposed ; namely, those
of Buffon, Lamarck, St. Hilaire, Darwin, and Nägeli.   Modern
criticism has highly differentiated, but not essentially altered
these hypothetical factors since they were originally conceived.
Darwin's 'survival of the fittest' we may alone regard as
absolutely demonstrated as a real factor, without committing
ourselves as to the 'origin of fitness.'   The third step is to
recognize that there may be an unknown factor or factors which
will cause quite as great surprise as Darwin's.   The feeling
that there is such first came to the writer in 1890 in considering the want of an explanation for the definite and apparently
purposeful character of certain variations.[1]   Since then a simi-

[1] *Op. cit.*, 1891.

lar feeling has been voiced by Romanes and others, and quite lately by Scott ;[1] but the most extreme expression of it has recently come from Driesch[2] in his implication that there is a factor not only unknown but unknowable !

Theoretically neither of these five hypotheses of the day excludes the others. They may all coöperate. The rôle which each plays, or the fate of each in the history of speculation largely or wholly depends upon the solution of the problem of the transmission or non-transmission of acquired variations and after all that has been written on this question this must be regarded by every impartial observer as still an open one.

We are far from finally testing or dismissing these old factors, but the reaction from speculation upon them is in itself a silent admission that we must reach out for some unknown quantity. If such does exist there is little hope that we shall discover it except by the most laborious research ; and while we may predict that conclusive evidence of its existence will be found in morphology, it is safe to add that the fortunate discoverer will be a physiologist.

THE ANALYSIS OF VARIATION.

After this introductory survey let us consider as another outcome of the controversy that Variation and the related branch of research, Experimental Evolution, are now in the foreground as the most important and hopeful of the many channels into which the inductive tests of known or unknown factors may be turned. Let us make an honorable exception of those reactionists, such as Bateson[3] and Weldon, who have instituted an exact investigation into the laws of Variation.

How shall the study of Variation be carried on ? I totally differ at the outset from Bateson in the standpoint taken in the introduction of his work, that the best method of starting such an investigation is in discarding the analysis which rests upon the experience as well as the more or less speculative basis of

---

[1] On Variations and Mutations. *Am. Jour. Sc.*, November, 1894.

[2] Analytische Theorie der Organischen Entwickelung. Leipsic, 1894.

[3] W. Bateson : Materials for the Study of Variation. London, 1894.

past research.   There is little clear insight to be gained by considering variations *en masse*, and in this lecture I shall put forth some reasons why this is the case as well as some principles which seem to be preliminary to an intelligent collection and arrangement of facts, upon the ground that a mere catalogue of facts will have no result.   Variation is to be regarded as one of the two modes or expressions of Heredity, or as the exponent of old hereditary forces developing under new or unstable conditions.   It stands in contrast not with Heredity, which includes it, but with Repetition as the exponent of old forces developing under old or stable conditions.   Nägeli ten years ago[1] laid stress upon this, as have latterly Weismann, Bateson, Hurst,[2] and others.   Nevertheless it is still widely misconceived.   Hurst even regards Variation as the oldest phenomenon — an error in the other extreme, for they are rather coincident phenomena — representing the stability or instability of development.   The broadest analysis we can make is that variations are divided by three planes — the plane of *time*, the plane of *cause*, and the plane of *fitness*.   This raises the three problems to be solved regarding each variation : when did the variation originate? what caused it to originate? is it or is it not adaptive?

The student of heredity, in connection with these three planes of analysis, has then to consider the modes of heredity as complementary or interacting, for as soon as a 'variation' recurs in several generations it is practically a 'repetition,' and the repetition principle is a frequent source of apparent but not real variation or departure in the offspring from parental or race type.   This relation becomes clear when we consider variations in man, as seen in anatomy and in Galton's studies of inheritance and expressed in the following table : —

[1] "Vererbung und Veränderung sind, wenn sie nach dem wahren Wesen der Organismen bestimmt werden, nur scheinbare Gegensätze." *Theorie der Abstammungslehre*, p. 541.

[2] Biological Theories.   I, The Nature of Heredity. *Natural Science*, vol. I, No. 7, September, 1892.   II, The Evolution of Heredity. *Natural Science*, vol. I, No. 8, October, 1892.

HEREDITY.

Repetition.	*Variation.*	
A. Retrogressive to present and past type.	*A.* Neutral both as regards present or future type. Including anomalies and abnormalities which are purely individual phenomena not in the path of evolution.	*B.* Progressive to future type.
(*a*) Repetition of parental type.		(*a*) Ontogenic variation from parental type in one or more characters.
(*b*) Regression to present race type usually in several characters(=Variation from present *parental* type).		(*b*) Ontogenic variation from present race in several characters (=a new sub-type).
(*c*) Reversion to past race type, usually in few or single characters (=Variation from present *race* type). *Palingenic Variation.*		(*c*) Phylogenic or constant variation towards future race type, in one or more characters, constituting a new 'Variety' (=Repetition of parental type). *Cenogenic Variation.*

The most profound gap in time is between ‘palingenic variations,’ springing from the past history of the individual, and ‘cenogenic variations,’ which have to do only with present and future history. The former embraces more than reversion. This table gives us only our first impression of this plane of time so lightly regarded by Bateson, if indeed discrimination is possible among data of the kind he has collected. The distinctive import of human anatomy [1] is that a comparison of the past and present habits of the race, or of the uses to which bones and muscles have been and are now being put, opens a possible analysis of variations both as regards their time of origin and as regards their fitness to past, present, or future uses; it is thus an inexhaustible mine for the philosophical study of variation — of which only the upper levels have been worked.[2] Beside the human organism there is no other within

[1] R. Wiedersheim : Bau des Menschen als Zeugniss seiner Vergangenheit. Freiburg, 1887.

[2] H. F. Osborn : Present Problems in Evolution and Heredity. The Cartwright Lectures. I. The Contemporary Evolution of Man, etc. Wm. Wood & Co., New York, 1891.

our reach admitting such exact analysis of variation in the
planes of time and fitness.   When, again, we connect human
anatomy as a field for the study of Variation with Galton's
researches, although his emphasis has been chiefly upon the
laws of Repetition, we begin to appreciate the far-reaching
importance of his inductions.   In contrast with those of Weismann they are based upon facts and will stand.   In the first
volume of these Marine Biological Laboratory lectures I went
into some detail to show how Galton bears upon the modern
evolution problem, so that here I may briefly recapitulate.   He
demonstrates two principles : First, that there must be some
strong progressive variational tendency in organisms to offset
the strongly retrogressive principle of Repetition wherever
the neutralizing or swamping effect of natural inter-breeding
is in force, as it virtually is for most anatomical characters of
the human race.   Second, he shows what has not been pointed
out in this connection before, that in natural inter-breeding
ontogenic or individual variations are conspicuous but in the
main temporary, while there is a strong undercurrent of phylogenic variations relatively inconspicuous and permanent.   Other
evidence supporting this latter principle comes out as we
proceed.

What is the value of a distinction between *ontogenic* and
phylogenic variations ?   It is this : it sets forth the widely
neglected initial problem of the *time of origin of a variation
in the life history of the individual*.   This is the first step in
experimentation upon variation, not only as it will afford
crucial evidence as to the factors of Buffon, Lamarck, and of
St. Hilaire, which hinge upon the inheritance of acquired variations, but in the coming days of exact research upon Variation
in general.   Let *ontogenic variation* — a term first used by
Brooks, I believe, although I cannot point out where — include
all deviations from type which have their cause in any stage
of individual development.   We are now beginning to fully
recognize that the causes of certain kinds of variation actually
can be traced to external influences upon certain stages of
growth or ontogeny, and that it will be possible ultimately to
determine these stages when this matter of time is established

by experiment.   Let *phylogenic variation* — a term first used by Nägeli [1] — include those departures from type which have become constant hereditary characters in certain phyletic series or even in a few generations.   While all phylogenic variations must originate in ontogeny or in some stage of individual development, certainly a very small proportion of the innumerable ontogenic variations which we find in the examination or measurement of any adult individual ever become phylogenic, or constitute more than ripples upon the surface of a tide.

This vital distinction has not been regarded hitherto.   The statistics of variation, as compiled by Darwin and lately by Wallace, Weldon, Bateson, and others, do not take into account that among phylogenic variations are others purely ontogenic springing up and disappearing during individual life, owing to causes connected solely with the disturbance of the typical action of the hereditary mechanism during ontogeny.   In other words, these writers have without discrimination based upon variations, which may be largely or wholly ontogenic and temporary, the important principles of 'Fortuitous Variation' of Darwin and of 'Discontinuous Variation' of Bateson, whereas it is only the laws of phylogenic variation which are of real bearing upon the problem of evolution.   Take as an illustration of this false method the wing measurements of birds given by Wallace.   Why may not these be largely cases of purely ontogenic variation due to influences of life habit or to some purely temporary disturbance of the hereditary basis?   Above all others, the Neo-Darwinians must reconsider their principle of 'fortuitous variation' which has been based upon data of miscellaneous ontogenic and phylogenic variations, because Neo-Darwinism is essentially and exclusively a theory of the survival of favorable phylogenic variations.

One aspect of the variation problem of to-day may, therefore, be stated thus : What is the cause, nature, and extent of

---

[1] Die Veränderung, die gewöhnlich der Vererbung gegenüber gestellt wird, steht nicht im Gegensatz zu dieser, sondern zur Constanz.   In diesem Sinne heisst eine Veränderung constant, wenn das Gewonnene dauernd behalten, und vergänglich, wenn es bald wieder preisgegeben wird.   Die constante oder die *phylogenetische Veränderung* . . . ist eigentlich nichts anderes als die Constitutionsänderung des Idioplasmas.   *Theorie der Abstammungslehre*, p. 277.

ontogenic variations in different stages of development, and under what circumstances do ontogenic variations become phylogenic ?

This brings us to an analysis of ontogenic variations in the *plane of time* as provisionally expressed in the following table : —

ORIGIN OF VARIATIONS DURING LIFE HISTORY.

A. Ontogenic Variations.	Theories of Causation.
(*a*) *Gonagenic, i.e.,* those arising in the germ-cells, including the ‘Blastogenic’ in part of Weismann, the ‘Primary Variations’ of Emery.	Theoretically connected with pathological, nutritive, chemico-physical, nervous influences, as implied by Kölliker and others, including doubtful phenomena of Xenia and Telegony.
(*b*) *Gamogenic, i.e.,* those arising during maturation and fertilization, including the ‘Blastogenic’ in part of Weismann, ‘Secondary,’ or ‘Weismannian variations’ of Emery.	Theoretically connected with influences named above, also with the combination of diverse ancestral characters, ‘Amphimixis’ of Weismann.
(*c*) *Embryogenic, i.e.,* those occurring during early cell division, including the ‘Blastogenic’ and ‘Somatogenic’ in part of Weismann.	Theoretically connected with extensive anomalies due to abnormal segmentation and other causes, as observed in the mechanical embryology of Roux, Driesch, Wilson, and others.
(*d*) *Somatogenic, i.e.,* those occurring during larval and later development after the formation of the germ-cells.	Connected with reactions between the hereditary developmental forces of the individual and the environment.

B. *Phylogenic Variations.*

Variations from type, originating in any of the above stages which become hereditary.

The above table illustrates limits which certainly should not be sharply drawn between the successive stages of ontogeny, although intermediate focal points of real distinction must exist. The four terms proposed are not in the sense of the ‘blastogenic’ and ‘somatogenic’ of Weismann, for there is no implication of his *petitio principi*, namely, of the separation of the hereditary substance or specific germ-plasm from the body-cells. Even before somatogenic separation has taken place we have little or no reason to believe that all the blastogenic, gonagenic, or gamogenic variations which may have arisen from various causes will become phylogenic.

If we carry our analysis into the ‘*plane of fitness*’ the first point which arises is whether variations are *normal*, including both cenogenic and palingenic variations, or *abnormal*, including teratological and other malformations. The terms 'fortuitous' and 'indefinite' as opposed to 'determinate' and 'definite' may be used apart from any theory, although they have sprung up as distinguishing two opposed views as to the principles of variation. 'Fortuity' strictly implies variation round an average mean, while 'definite' is not the necessary equivalent of adaptive, but simply implies progressive or phylogenic variation in one direction which Waagen and Scott have termed "Mutation." Bateson's terms 'Continuous' and 'Discontinuous' are useful as distinguishing gradual from sudden ontogenic variation.

In general our five working hypotheses as to the factors of evolution are theoretically related to the time stages of Variation as seen in the following table : —

$$
\text{Buffon's} \left\{ \begin{array}{l} \textit{Ontogenic} \\ a \text{ Gonagenic} \\ b \text{ Gamogenic} \\ \text{ Allei} \\ c \text{ Embryogenic} \left.\right\} \text{St. Hilaire's} \\ d \text{ Somatogenic} \left.\right\} \\ \text{Darwin's} \left\{ \textit{Phylogenic} \right. \end{array} \right\} \text{Lamarck's}
$$

I again call attention to the fact that Neo-Darwinism has hitherto presupposed and practically assumed 'fortuitous phylogenic variation' as its basis, for it is solely related with the selection of those ontogenic variations which are also phylogenic. Neo-Lamarckism, on the other hand, is solely connected with inheritable 'somatogenic' variation. Buffon's factor of the 'direct action of the environment' plays upon all four ontogenic stages, and both theoretically and as observed by experiment, produces profound ontogenic variations ; the question is, under what circumstances do such ontogenic variations in each of the four stages become phylogenic? This factor would be partly but not wholly set aside by proof that somatogenic variations are not inherited. St. Hilaire's factor of the action of environment upon early stages of development would result in purely fortuitous variations, and, as he himself clearly perceived, would require Selection to give it an adaptive direction.   Nägeli's factor, on the other hand, assumes definite but not necessarily adaptive 'phylogenic' variation — his views have been very generally misconceived on these points — and, as he pointed out, his factor would also require Selection to determine which of the definite lines of growth were adaptive.

It seems necessary to thus clearly state the relations of the time stages of variation to each of the five factors, in order to show the decisive bearings our future exact research will have upon them.   For example, the proof that variation is either 'definite' or that it is 'adaptive' prior to or independently of Selection, will constitute conclusive disproof not of Darwin's theory but of Neo-Darwinism.   The fate of Lamarckism, on the other hand, depends upon the demonstration that phylogenic variation is not only 'definite' and 'adaptive' but that it is anticipated by corresponding somatogenic variation.

A review of recent thought upon the variation problem shows that these life stages are becoming generally recognized. I shall pass by Lamarck's and Darwin's factors which are so thoroughly understood and speak only of the other three.

BUFFON'S FACTOR IN VARIATION.

As regards Buffon's factor, which is the most comprehensive of all, we know that Spencer and Weismann both assumed that the direct action of the environment was primarily a factor of evolution.   Weismann first regarded this solely as the proto-zoan source of Variation, but has recently given it a wider play in the action of environment upon the germ-cells as a cause not of definite variation but of variability.   The line of research upon the dynamic action of environment in its influence upon somatogenic variation followed by Hyatt, Dall, and others, is paralleled in the more recent specula-tion connecting the environment directly with the gonagenic and gamogenic stages, initiated by Virchow,[1] Kölliker,[2]

---

[1] R. Virchow: Descendenz und Pathologie. *Virchow's Archiv*, CIII, 1886, pp. 1–15, 205–215, 413–437. Ueber den Transformismus. *Archiv f. Anthropologie*, 1889, p. 1.

[2] Kölliker: Das Karyoplasma und die Vererbung. *Zeitschr. f. wissenschaftl.*

Ziegler,[1] Sutton, and others. In a similar vein are the suggestions of Geddes, while those of Gerlach and Ryder direct our attention mainly to mechanical alterations in the embryonic stages of development. Botanists such as Vines, Detmer, and Hoffmann have pointed to the influence of environment upon gonagenic variation. Experiments of a general character resulting principally in embryogenic and somatogenic variation have been recently carried on by Cunningham, Agassiz, and others, as illustrating the direct action of the environment. Followers of Buffon's factor are also more or less identified with Lamarckism. The distinction is mainly expressed in the terms 'otagenic' and 'kinetogenic' of Ryder; for under Buffon's factor the organism is passive, while under Lamarck's it is active. Among others who have supported Buffon's principle are Packard, Eimer, Cunningham, Ryder, and Dall.

This literature and so-called 'evidence' upon Buffon's factor exhibits the greatest confusion of interpretation, and demonstrates that our conceptions first, as regards heredity, second, as regards variation under a changed environment, require thorough recasting.[2] First as regards evolution in relation to heredity. The reversion phenomena as seen in human anatomy wholly set aside Weismann's conception of evolution as the selection of favorable and the elimination of unfavorable hereditary variations; in other words, of selection acting directly upon the germ-plasm. These phenomena indicate rather that the direct process is not one of elimination but of suppression from the later stages of ontogeny, and that only after an enormous interval of time does actual elimination occur. Abnormal nervous conditions such as seen in Anencephaly are accompanied by the revival of a large number of latent characters. In Galton's language, patent characters become latent in the course of evolution.

Zoologie, 1886. Eröffnungsrede der ersten Versammlung der Anatomischen Gesellschaft in Leipzig. *Anat. Anzeiger*, II, 1887.

[1] Ernst Ziegler: Die neuesten Arbeiten über Vererbung und Abstammungslehre und ihre Bedeutung für die Pathologie. Tübingen.

[2] J. T. Cunningham: The Problem of Variation. *Natural Science*, vol. III, pp. 282–287. Also, Researches on the Coloration of the Skins of Flat-Fishes. *Jour. Mar. Biol. Assoc.*, May, 1893. (See also *Trans. Roy. Soc.*, 1892–3.)

In Weismann's language, on the other hand, in explanation of dimorphism in hymenoptera and other types, there are certain sets of biophors corresponding to certain possibilities of adult development. Apply this to the celebrated case of the flat-fishes and the remarkable results recently obtained by Agassiz, Filhol, and Giard in artificially producing more or less symmetrical flat-fishes by retaining the young near the surface. Weismann's interpretation of the evolution of flat-fishes has always been that it was by the selection of asymmetrical and elimination of symmetrical 'determinants.'   In the light of these experiments he must now recast this explanation by saying that the flat-fishes have kept in reserve a set of symmetrical 'determinants' since the period when our first record of the asymmetrical type appears or about three million years !

This attack upon the speculations of one writer is a digression. What I really wish to bring out is the necessity of a far more critical analysis of the various kinds of evidence for Buffon's factor.   This necessity may be illustrated by the different interpretations of color change in direct response to changed environment.

The most significant experiments upon color are those of Cunningham upon the flat-fishes.   He has proved that during the early metamorphosis of young flat-fishes, when pigment is still present on both sides, the action of reflected light does not prevent the disappearance of this pigment upon the side which is turned towards the bottom, so that the color passes rapidly through a retrograde development; but prolonged exposure to the light upon the lower side causes the pigment to *reappear*, and upon its reappearance the pigment spots are in all respects similar to those normally present upon the upper side of the fish.   It is very important not to confuse these results, of deep interest as they are, with those obtained where the environment is new in the historic experience of the organism.   Experiments upon color, therefore, afford a marked illustration of the necessity of drawing a sharp distinction between cenogenic and palingenic variations.   We have, in many cases, been mistaking repetitions of ancient types of structure for newly acquired structures.   When the pale *Proteus* is taken from the Austrian caves, placed in the sunlight, and in the course of a month becomes darkly pigmented, there are two interpretations of this pigmentation : either that we have revived a latent character, or that we have created a new character. The latter interpretation can alone be taken as a proof of Buffon's factor when it is found to be followed by hereditary transmission.

Poulton,[1] as a supporter of Neo-Darwinism, takes this view, in reply to Beddard and Bateson, and as an induction from his beautiful and exact experiments upon the coloring of lepidopterous larvae. After producing the most widely various colorings and markings by surrounding the larvae during ontogeny with objects of different colors, he urges that the changes thus directly produced simply revert to adaptations to former conditions of life, in other words, that they are palingenic. Whether this interpretation is correct or not, Poulton proves that, no matter how stable certain hereditary characters may appear to be, repetition in ontogeny depends upon repetition in environment, and that there are wide degrees of ontogenic variations which do not become phylogenic at least in several successive generations.

From many other analogous researches we gather the following principle to which far too little attention has been paid in the study of the phenomena of variation in their bearing upon the factors of evolution : *It is that ontogenic repetition depends largely upon repetition in environment and life habit, while ontogenic variation is connected with variation in environment and life habit.* If the environment be changed to an ancient one, then ontogenic variations tend to regression or reversion (*i.e.*, palingeny) or practically to repetition of an ancient type. It is necessary to state clearly that there is practically conclusive evidence for such a principle, not only in the later stages of development, as in the respiratory metamorphosis of the Amphibia, but extending back to very much earlier stages

[1] E. B. Poulton : Further experiments upon the color-relation between certain lepidopterous larvae, pupae, cocoons, and imagines and their surroundings. *Trans. Ent. Soc.*, pt. IV, p. 293. London, 1892. (Contains a reply to Beddard and Bateson.)

than we have hitherto suspected.   Thus a vast amount of evidence which has been brought forward as proof of Buffon's factor, *i.e.*, of the direct action of environment in producing definite and adaptive ontogenic variations is in reality in many cases no proof at all.

Having thus eliminated errors of interpretation, the great question still remains as to what happens when the environment is a wholly new one in the historical experience of the organism.   Do the ontogenic variations exhibit a new direction?   Is this direction adaptive, *i.e.*, towards progressive adaptation?   What relations have such new conditions to the hereditary potencies of the germ-cells?

Out of all actual researches it becomes clear that experimentation can henceforth be separately directed upon the four stages of development, and that it will be possible in some degree to draw such lines of separation.   New mechanical and chemical influences can be applied in each stage and withdrawn in the subsequent stages, the difficulty being to reach the extreme point where a profound influence is exerted without interfering with the reproductive functions.

One effect of new environment upon the gonagenic, gamogenic, and embryogenic stages will be *saltation*.   Ryder[1] has recently treated this in a most suggestive manner in discussing the origin of Japanese gold-fish.   Turning to St. Hilaire's hypothesis, we find he had in mind embryogenic variation mainly traceable to respiratory and chemical changes.   Virchow extends the cause of sudden change further back to chemico-physical influences upon the germ-cells.   The causes and modes of sudden development arising from whatever ontogenic stage demand the most careful investigation, chiefly in their bearing upon the relation of ontogenic to phylogenic variation. Galton has discussed the subject objectively under the head of 'Stability of Sports,' and Emery, under the head of 'Primary Variations, has supported Galton's observation that such salta-

---

[1] The inheritance of modifications due to disturbance of the early stages of development, especially in the Japanese domesticated races of gold-carp.   *Proc. Acad. Nat. Sc. Phila.*, 1893, p. 75.

tions often exhibit a strong capacity for inheritance.   Bateson reaches in the conclusion of his work a modified form of St. Hilaire's factor of saltatory evolution, and believes that species have largely originated by 'discontinuity' of variation or the sudden accession of new characters from unknown causes, concluding that all inquiry into the causes of variation is premature.   The materials he has brought together are of the greatest value, and he has already been able to throw in doubt many current beliefs, such as that variability is greater in domestic than in wild animals.   His interpretation of these materials is, as we have seen, weakened, so far as it bears on our search for the evolution factors, by the fact that from the nature of most of his evidence he cannot discriminate between ontogenic and phylogenic variation : moreover, he discards any attempt to discriminate between palingenic and cenogenic variations.   This lack of analysis leads him into what appears to be an entirely erroneous induction, for the principle of discontinuity is opposed by strong evidence for continuous and definite phylogenic variation as observed in actual phyletic series.

Nägeli's Factor and Phylogenic Variation.

Nägeli's factor [1] introduces us to an entirely distinct territory — to the opposite extreme from saltation.   It is one we can no longer set aside as transcendental because of the strong likeness it bears at first sight to the internal perfecting principle of Aristotle.   It is supported in a guarded manner by Kölliker and Ziegler.   It contains the large element of truth that the trend of variation and hence of evolution is predestined by the constitution of the organism ; that is, granted a certain hereditary constitution and an environment favoring its development, this development will exhibit certain definite directions, which when reaching a survival value will be acted upon by selection. I have recently [2] described as the '*potential of similar varia-*

[1] C. v. Nägeli : Mechanisch-physiologische Theorie der Abstammungslehre. München und Leipzig, 1884.

[2] Rise of the Mammalia in North America. *Stud. Biol. Dept. Columbia College*, vol. I, No. 2, September, 1893.

tion' an evolution principle which seems to be well supported by palaeontological evidence.   It is this: while the environment and the activity of the organism may supply the stimuli in some manner unknown to us, definite tendencies of variation spring from certain very remote ancestral causes; for example, in the middle Miocene the molar teeth of the horse and the rhinoceros began to exhibit similar variations; when these are traced back to the embryonic and also to the ancestral stages of tooth development of an early geological period, we discover that the six cusps of the Eocene crown, repeated to-day in the embryonic development of the jaw, were also the centers of phylogenic variation; these centers seem to have predetermined at what points certain new structures would appear after these two lines of ungulates had been separated by an immense interval of time.   In other words, upper Miocene variation was conditioned by the structure of a lower Eocene ancestral type.

This is the proper place to recall a kindred conception of Variation which has been in the minds of many, and has been clearly formulated it appears by Waagen.   It is of Variation so inconspicuous and so slight that it can only be recognized as such when we place side by side two individuals separated by a long series of generations.[1]   Mark the contrast with the extreme of St. Hilaire's saltatory evolution; or again, the contrast with Darwin's and Weismann's conception of Variations, not, it is true, of a saltatory character, but as sufficiently important and conspicuous to become factors in the survival of the organism.   This conception of 'phylogenic variation,' as we have seen, is consistent with the application of Galton's principles to human evolution, but it finds its strongest support in palaeontology, and is the unconscious motive of dissent on the part of all palaeontologists, so far as I know their opinions, independently working in all parts of the world, to the fortuitous Variation and Selection theory.

---

[1] This was brought out by the writer in his Oxford paper. See *Nature*, August 30, 1894, p. 435. It has recently been independently stated with great clearness by Scott in his article Variations and Mutations. *American Journal of Science*, November, 1894. Scott, following Waagen, revives the term 'mutation' for what Nägeli has termed 'phylogenic variation.'

Our palaeontological series are unique in being phyletic series. They exhibit no evidences of fortuity in the main lines of evolution. New structures arise by infinitesimal beginnings at definite points. In their first stages they have no 'utilitarian' or 'survival' value. They increase in size in successive generations until they reach a stage of usefulness. In many cases they first rise at points which have been in maximum use, thus appearing to support the kinetogenesis theory. In extensive fossil series we also find evidence of anomalous or neutral variations, such as Bateson has brought together, but these are aside from the main lines of evolution. They present no evidence for the Neo-Darwinian principle of the accumulation of adaptive variations out of the fortuitous play around a mean of adaptive and inadaptive characters, but they present strong evidence of the Darwinian principle of the survival of the fittest. The main trend of evolution is direct and definite throughout, according to certain unknown laws and not according to fortuity. This principle of progressive adaptation may be regarded as inductively established by careful studies of the evolution of the teeth and the skeleton. Its bearing upon Lamarck's factor of the transmission of somatogenic variation was pointed out by myself in 1889; it does not positively demonstrate Lamarck's factor because it leaves open the possible working of some other factor at present unknown, and Lamarck's factor is also inadequate; but it positively sets aside Darwin's factor as *universal* in the origin of adaptations and as a consequence 'the all-sufficiency of Natural Selection.' If Lamarck's factor is disproved, in other ways, it leaves us *in vacuo* so far as a working hypothesis is concerned.

The conclusions which Hyatt, Dall, Williams, Buckman, Lang, and Würtemberger have reached among invertebrates are independently paralleled by those of Cope, Ryder, Baur, Scott,[1] the writer, and many other morphologists. The same general philosophical interpretation of evolution is now independently announced from an entirely different field of work by Driesch. We may waive our applications of these facts to theories, but let us not turn our backs to the facts themselves!

[1] W. B. Scott: On Some of the Factors in the Evolution of the Mammalia. *Journ. of Morphology*, vol. V, 1891, p. 378.

THE OUTLOOK FOR INDUCTION.

The problem just raised is the main one. No longer misled by palingenic variation under revival of an ancient environment, let us set ourselves rigidly to the analysis and investigation of the responses of the organism to new environment, in all four stages of development. Are these responses adaptive? Is there a teleological mechanism in living matter as Pflüger[1] has expressed it? Is this mechanism in the adult reflected in the germ?

One most hopeful outlook is in Experimental Evolution. Bacon in his *Nova Atlantis* three centuries ago projected an institute for such experiments, which when it finally materializes should be known as the Baconian Institute. The late Mr. Romanes proposed to establish such a station at Oxford, and went so far as to institute an important series of private experiments, which were unfortunately interrupted by his death. What we wish to ascertain is, whether new ontogenic variations become phylogenic, and how much time this requires.

The conditions of a crucial experiment may be stated as follows: An organism A, with an environment or habit A, is transferred to environment or habit B, and after one or more generations exhibits variations B; this organism is then re-transferred to environment or habit A, and if it still exhibits, even for a single generation, or transitorily, any of the variations B, the experiment is a demonstration of the inheritance of ontogenic variations. These are virtually the conditions rightly demanded by Neo-Darwinians for an absolute demonstration, either of Lamarck's or Buffon's principle of the inheritance of embryogenic or somatogenic variation. There is no record that such conditions have as yet been fulfilled, for hitherto organisms have been simply retained in a new environment, and the profound modifications which are exhibited may simply be the exponents of an hereditary mechanism acting under the influence of new forces. Such experiments will probably require an extended period of time, for we learn from palaeontology, as well as from palingenic variation,

---

[1] Pflüger: Die teleologischen Mechanik der lebenden Natur. Bonn, 1877.

that phylogenic inheritance is extremely slow in a state of nature.

It is desirable to establish non-infectious experimentation involving the conditions named above, mainly as a test of Lamarck's factor. Varigny has also proposed a crucial experimental series mainly upon Buffon's factor. His volume upon *Experimental Evolution* is an invaluable review, especially of French researches in experimental transformism. Much of this is in the line brought together some years ago by Semper in his *Animal Life.* Varigny draws a valid distinction between morphological variation and physiological variation, including under the latter internal chemical and constitutional differences which are not displayed in structure but must underlie all reactions. Under the head of what I have called Gonagenic Variation, the author discusses the work of Gautier[1] upon the influence of previous fertilization in plants as well as upon the chemistry of plants in connection with color variation. He adds to the observations of Yung and Born other studies upon sex determination. He describes the experimental teratogeny or embryonic variation of Dareste, Fallou, and later observers.

Throughout Varigny's volume it is nevertheless evident that none of the studies upon Ontogenic Variation hitherto have been specifically directed to the vital problem, as they must be in the future. Varigny makes a useful suggestion as to the importance of imitating natural conditions in experimental work, but he fails to emphasize the importance of the tests set forth above in order to ascertain whether the acquired modifications have actually been impressed upon the hereditary mechanism or merely upon the various stages of ontogeny.

The general conclusion we reach from a survey of the whole field is, that for Buffon's and Lamarck's factors we have no theory of Heredity, while the original Darwin factor, or Neo-Darwinism, offers an inadequate explanation of Evolution. If acquired variations are transmitted, there must be therefore

[1] Armand Gautier: Du Mécanisme de la Variation des Êtres vivants. (Hommage à Monsieur Chevreul à l'Occasion de son Centennaire. F. Alcan. Paris. 1886.)

some unknown principle in Heredity; if they are not transmitted, there must be some unknown factor in Evolution.

As regards Selection, we find more than the theoretical objections advanced by Spencer and others. Neo-Darwinism centers upon the principles of fortuitous variation, utility, and selection as universal. In complete fossil series it is demonstrated that these three principles, however important, are not universal. Certain new adaptive structures arise gradually, according to certain definite laws, and not by fortuity.

Lamarck's and Buffon's factors afford at present only a partial explanation of these definite phylogenic variations, even if the transmission of acquired variations be granted. Nägeli's factor of certain constitutional lines of variation finds considerable verification in fossil series as a principle of determinate variation, but not as a general internal perfecting tendency. St. Hilaire's factor of occasional saltatory evolution by sudden modification of the hereditary mechanism is established, but not as yet understood, although we are perhaps approaching an explanation through experimental embryology.

Our standpoint towards Variation in relation to all the Factors requires thorough reconsideration. The Darwinian law of Fortuity and the Buffon law of the direct action of Environment, have hitherto been inductions from variations which may be largely ontogenic and transitory. They both require confirmation on data of phylogenic variation. As for Lamarck's factor, the evidence seems to be conclusive that somatogenic variation is largely adaptive; but it remains to be proved that phylogenic variations as observed in human anatomy and in palaeontology are invariably anticipated by corresponding changes in the individual, in other words, that the definite current of variation is guided by the inheritance of individual reactions.

Another consideration is, that individual Variation may play a far less conspicuous rôle than we have assigned to it; in other words, that many of the most important changes in successive generations are so gradual as to be entirely inconspicuous in a single generation.

Our conception of the mechanism or physical basis of

Heredity is also to be made much clearer by a series of experiments directed to palingenic variation, in order to ascertain how far the revival of an ancient environment arouses latent hereditary forces. The experiments already well advanced by Cunningham, Agassiz, and Poulton indicate that *progressive inheritance is rather a process of substitution of certain characters and potentialities than the actual elimination implied by Weismann.*

My last word is, that we are entering the threshold of the Evolution problem, instead of standing within the portals. The hardest tasks lie before us, not behind us, and their solution will carry us well into the twentieth century.

3

WILHELM ROUX
1850–1924

Because of his aggressive style and his journal, *Archiv für Entwickelungs-mechanik der Organismen*, Roux had become a familiar name to biologists by 1894. After completing his graduate work at Jena, Roux began a career at Breslau which was to span the decade 1879–1889. During his stay there, Roux, Eduard Pflüger, and Gustav Born each pursued parallel study of frogs' eggs. Roux's theoretical work, *Der Kampf der Theile in Organismus* (1881), provided a speculative view of natural selection operating within the organism. Roux maintained that the mechanistic action of the struggles brought by natural selection would account for growth and development. He followed that provocative study with more experimental analysis of development, turning by 1883 to frogs in particular.

Pflüger's experimental work on the effects of gravitation on cleavage inaugurated a series of experimental studies of frog development and cleavage. In the next few years, Roux and others pursued experimental disruption of cleavage, and by 1887 found that puncturing one of the two blastomeres produced only a half blastula. For Roux, that result provided sufficient evidence for what he termed "mosaic development," according to which each cell of a developing organism differentiates independently and follows the dictates of the mosaic produced by qualitative division during cleavage. He further advocated that only experimentation could produce valid results for developmental study. That program for *Entwickelungsmechanik* found expression in his articles in his journal, which began publication in 1894 and received considerable attention at the MBL. The essay that follows is a classic example of Roux's self-confident—indeed, bombastic—style, and was a major influence on the development of modern biology. Wheeler evidently became aware of the enthusiasm for Roux's experimental program at Naples and produced this translation upon his return to the MBL in 1894.

WILHELM ROUX.

THE PROBLEMS, METHODS, AND SCOPE OF DEVELOPMENTAL MECHANICS

An Introduction to the " Archiv für Entwickelungsmechanik der Organismen."

WILHELM ROUX

[Translated from the German by WILLIAM MORTON WHEELER.] [1]

I. *The Problems of Developmental Mechanics.*

DEVELOPMENTAL mechanics or causal morphology of organisms, to the service of which these " Archives " are devoted, *is the doctrine of the causes of organic forms, and hence the doctrine of the causes of the origin, maintenance, and involution (Rückbildung) of these forms.*

Internal and external *form* represents the most essential attribute of the organism in so far as form conditions the *special manifestation* of life, to which the genesis of this form itself in turn appertains.

The term "mechanics of development," to designate the causal doctrine of this whole subject, is employed in accordance with the principle *a potiori fit denominatio,* for the *evolution* of organic form comprises the main processes and implies the principal problems of organic formative operations.

[1] The translation of this philosophical essay has been attended with not a few difficulties. Besides the difficulties resulting from the great compactness of Professor Roux's style, there are others, not the least of which are the great conciseness of meaning with which all the terms are used, and the often very delicate qualifications of the leading ideas in the various paragraphs and sentences. I believe that I have rendered the ideas truthfully in the main, but I fear that it has been at the expense of a somewhat forced and unnatural construction in many of my sentences. — W. M. W.

In accordance with Spinoza's and Kant's definition of mechanism, every phenomenon underlying causality is designated as a *mechanical phenomenon;* hence the science of the same may be called mechanics. Since only phenomena underlying causality are capable of investigation, and hence alone may be made the subject of an *exact science,* and since the production of *form* constitutes the essential feature of development, it is quite permissible to call the science of the causes of form developmental mechanics.

Since, moreover, physics and chemistry reduce all phenomena, even those which appear to be most diverse, *e.g.,* magnetic, electrical, optical, and chemical phenomena, to movements of parts, or attempt such a reduction, the older more restricted concept of mechanics in the physicist's sense as the causal doctrine of the movements of masses, has been extended to coincide with the philosophical concept of mechanism, comprising as it does all causally conditioned phenomena, so that the words "developmental mechanics" agree with the more recent concepts of physics and chemistry, and may be taken to designate the doctrine of all formative phenomena.

Inasmuch as we call the *causes* of every phenomenon *forces* or *energies,* we may designate as the general problem of developmental mechanics *the ascertainment of the formative forces or energies.* In so far, however, as forces or energies are only known to us by their *effects, i.e.,* every kind of force by its specific *mode of operating,* the problem may be defined as the *ascertainment of the formative modi operandi.*

In accordance with this statement, a *general,* not quantitative, but in the first instance, merely *qualitative causal explanation* will always consist in tracing back a particular phenomenon to *modi operandi of more general validity, i.e.,* to such as operate *constantly,* also in many other processes, and hence under the same conditions, at all times and in all places, and in the same manner. Such modes of operating may be called "constants of operation" ("Wirkungsbeständigkeiten").

These constant modi operandi which follow from the properties of the components and hence of necessity, — these so-called *uniformities of nature,* — are usually called "*natural laws.*"

Accepting this latter term, the task of developmental mechanics would be the reduction of the formative processes of development to the natural laws which underlie them.

It is, however, preferable, at least in those cases to which the expression *constant mode of operating* is more applicable, to employ this phrase instead of the term *natural law*, which is based upon anthropomorphic conceptions of nature. It behooves us, especially when entering on a new and extensive field of investigation, beset with quite special difficulties, to call *the thing to be sought by its own name*, instead of employing an expression which is foreign to its nature.

Since, moreover, all the *modi operandi* underlying causality, and hence all *modi operandi* which may become the subject of our investigation, are "constant or uniform," this adjective may generally be omitted, and it is sufficient to say simply *modi operandi*, instead of natural laws. Instead of the "law" of the refraction of light we may also speak of the *modus operandi* of refraction; instead of the "laws" of functional adaptation let us say the *modi operandi* of functional adaptation, *e.g.*, of the muscles. This designation at the same time renders impossible in Biology one widespread, incorrect usage of the term "law," viz., the use of the term to designate *facts* or *results* instead of *operations*, as, for instance, in the current expression "Bell's law." When we attempt to use, instead, Bell's *modus operandi*, it becomes at once apparent that this term is inapplicable to the "fact" of the motor nature of the anterior, and the (supposed) purely sensory nature of the posterior nerve-roots.

If, furthermore, we define *the general task of developmental mechanics* so that it shall include the fewest mysterious concepts, and hence in a way which is simplest and most compatible with the immediate method of procedure, *we must reduce the processes of organic formation to the fewest and simplest modi operandi*. This, of course, implies that for each of these modes the *simplest expression* is to be sought.

All *operating*, and hence also its product, all *operation*, has at least two causes or components, since in last analysis nothing can change its condition of itself.

Development is a change and must, therefore, always depend on several components, and hence on *combinations of causes or energies.* More accurately speaking, we understand by development the *production of multiformity.* The latter results from every operation, from every combination of energies, at least during and for a short time after the duration of the operation ; and its origin depends on the *unequal distribution* of energy during its transmission, *e.g.,* in pressure on a body, in heating or electrifying an object, in the radiation of light-rays, etc. It is, therefore, unnecessary *in principle* to postulate *specific energies of development ;* this, however, does not preclude a possible participation at the same time of special components, as, *e.g.,* the energies of growth, in producing *formative* diversity during particular phases of organic development.

Organic development consists in the production of perceptible, *typically constituted* diversity. If we look aside in this place from the conditions of perception (1), *typical combinations of causes or energies are indispensable to the origin of " typical diversity."* For the specifically *constituted* nature of this diversity, specific *form-producing combinations of causes are required,* and these represent the just-mentioned "formative components." Now if these formative components be forthcoming in a perfectly typical manner, in kind, magnitude, and arrangement, it is self-evident that in the absence of disturbance from without, the constructive diversity produced by these components must be perfectly typical.

Accordingly, in any given case, we must trace back each individual formative process to the special combination of energies by which it is conditioned, or, in other words, to its *modi operandi ;* and each of these *modi operandi* must be ascertained with respect to *place, time, direction, magnitude,* and *quality.* Or, inversely, we may endeavor to determine *in the individual structure the special part which is performed by every modus operandi known to participate in the development of the organism.*

These *modi operandi,* to which we reduce organic formative processes, and hence also the energies which condition them, may be identical with those which underlie inorganic or physico-chemical processes.

Since it is not the task of the biologist, as such, to investi-gate and to subject the components of *inorganic* phenomena to an analysis further than that undertaken by physicists and chemists, we may accept these components as given, and may designate them, so far as they are concerned in organic opera-tions, as "SIMPLE COMPONENTS," no matter how problematic their nature may be, and even if sooner or later they should be still further dissociated by physicists and chemists. When this is accomplished, we shall make use of these further, still simpler components.

Besides the endeavor to ascertain such " simple components," the lines of research in developmental mechanics must from the start be guided by the conviction that *organic structure is mainly due to the operation of components which at present are so complicated as to exceed the limits of our observation.* For these I have suggested the term "COMPLEX COMPONENTS" (2). Although according to our immediate conception of the matter, even these components depend in the last instance on inorganic *modi operandi*, nevertheless *the complexity of their composition lends them attributes which often differ so widely from those of inorganic modi operandi* that they are not only very *dissimilar* but even *appear to contradict* in part the functions of these same inorganic *modi operandi*. This is the case with the non-exosmosis of salts from living fish-eggs in water, the non-desic-cation of small living insects in the sunlight; whereas after death, these organisms, in the former instance suffer diosmosis, in the latter desiccation ; another instance is the pouring of a glandular secretion into a cavity which is under higher pressure than that which obtains in the blood capillaries of the gland. These processes show that in the former instances the salt or the water is not in a free state, but fixed and operant (beschäftigt), whereas in the last instance we are dealing with specific active functions carried out with commensurate expense of energy on the part of the epithelial cells.

It must, therefore, be our next most important task to ascer-tain these components, *which, though complex, are nevertheless*

alike constant and always alike operant under like conditions,
i.e., to reduce organic formation to such *modi operandi* as are
constant, albeit in themselves not understood.

Every "complex component" thus represents merely the
effect, the resultant of inappreciable individual effects. From
such complex components result most of the formative processes
which we perceive; it is our task, therefore, to analyze the
chaos of internal operations into the least possible number of
such *modi operandi*.

In the first place the elementary cell-functions are such
"COMPLEX COMPONENTS": *assimilation, dissimilation* (katabo-
lism) the *self-movement* of the cell in general, the *self-division*
of the cell as a definite coördination of self-movements; to these
we may add the *typical formal self-constructivity* and the *qualitative self-differentiation* of the cell as still more highly complicated effects.

On the other hand, the *growth in mass* of cells probably represents only the resultants of simultaneously occurring processes of assimilation and dissimilation; and the same may
hold good with reference to external pressure when the cell
decreases in mass. *Local growth,* however, besides depending
on a growth in mass of the cells of a given area may also
depend on the immigration of cells, and hence on other complex components, such as *chemiotropism* and *cytotropism* (3).
On the other hand, exclusively "*dimensional growth*" (4) of an
area may depend on the active metamorphosis of cells. Further
complex components which also determine the direction of
movements in unicellular or multicellular organisms are *galvano-,*
helio-, hydro-, and *thigmotropism.*

The *directive effect of the "form"* of the cleavage-cell prior
to its histological differentiation *on the position of the nuclear
spindle,* viz., the adjusting of the spindle to coincide with the
longest axis that can be drawn through the center of mass of
the protoplasm (5); *the trophic effect of functional stimuli* (to
which all the extraordinarily diverse phenomena of functional
adaptation are reducible) (6); *the trophic effect of ganglion cells
on their nerve-fibres* and corresponding end-organs — all these
are further complex components which are already established,

and through which many formations are attained.   The *effect of increased blood-supply* on the increase of connective tissue in the affected parts is another instance.

These complex components seem relatively simple in comparison with others which must be postulated before we can begin the analysis of many structures.

As an example of these the following may be formulated, if only provisionally; for if we never have the courage to begin we shall never escape from our ignorance.

The cells of all tubular and acinous glands have a *bipolar* differentiation; they have a *basal surface* which serves to take up nutriment from the adjoining capillaries, and opposite this a *secreting surface;* both surfaces are separated by the *whole* diameter of the cell; the remaining surfaces are merely *surfaces of contact* with the neighboring cells.   Metabolism is carried on in the direction of the axis uniting the polar surfaces, which direction is usually that of the greatest dimension.

The arrangement of the cells in lobules in the fully developed mammalian liver, which is a *reticular gland* with the narrowest possible meshes, viz., meshes only the breadth of a single cell in diameter, causes the cells to be *multipolar* in the above sense, for each cell has several nutriment-absorbing and several secreting surfaces.   The secreting and nutriment-absorbing surfaces are removed from one another by only *half* the cell-diameter.   The lobular structure composed of these cells represents, so far as its form is concerned, merely a cast of the interstices between the meshes of the network of tubular blood capillaries.

Inasmuch as the lobular structure, molded as it is on the blood capillaries, presupposes the small-meshed reticular type and this in turn the multipolarity of the liver cells, we may regard as the *primitive* factor in all these deviations from the tubular type of other glands, the change in the polarity of the liver cells, and we may say accordingly : The transformation of the tubular type, which is also present at first in the mammalian liver, into the definitive lobular type is the consequence of the differential change of the original bipolar nature of the liver cells to a multipolar nature, or ; *the multipolar differenti-*

*ation of the liver cells conditions or effects the transformation of
these cells from the tubular to the lobular type*, whereby the
lobule for purposes of best nutrition accommodates itself intimately to the tubular blood-capillaries.

All these fixed "constant *modi operandi*" of organic formative processes must be still further determined with respect
to their place and the time, direction, and extent of their participation in the special structures of organisms, and with
respect to their mode of operating.

In the first place we shall have to ascertain a great number
of such constant *modi operandi*, and all of these must then be
further decomposed into simpler and more widely distributed
complex components. In this undertaking it will probably be
frequently possible *to disentangle a simple component from
among the complex components.*

The immediate result of this undertaking, as in every
analysis, will be complication instead of simplification, since
apparently simple processes will often be separated into two
or more components. The simplifying effect of the analysis
will only appear after it has been extended to *many* processes
with the result of repeatedly finding the *same* components.

This simplifying effect is already apparent: all the extremely
diverse structures of multicellular organisms may be traced
back to the few *modi operandi* of cell-growth, of cell-eva-
nescence (Zellenschwund), cell-division, cell-migration, active
cell-formation, cell-elimination, and the qualitative metamorphosis of cells; certainly, in appearance at least, a very simple
derivation. But the infinitely more difficult problem remains
not only to ascertain the special rôle which each of these processes performs in the individual structure, but also to decom-
pose these complex components themselves into more and more
subordinate components.

And notwithstanding such apparent simplicity, the formative
causes in each higher aggregation of living units may differ in
part from the formative causes of a lower order, as, *e.g.*, formative modes which belong to the independently existing lower
units, such as the Protista, may be absent in the higher state
of aggregation from the corresponding units, viz., the cells of

a multicellular organism; while at the same time, new effects are produced which are peculiar to the *higher unit* in question and which would naturally depend on the *reciprocal* operations of the lower constituents. After *ascertaining the formative functions of each such unit*, the *modi operandi* on which these functions depend must be established by themselves; this holds good in the case of the *lowest independent parts* of the cell (7): the *isoplassons, autokineons, automerizons, idioplassons*, and the parts which they constitute, *the nucleus, centrosome*, and *protoplasm*. It also holds good in the case of the entire cells themselves, of the tissues, organs, and the organism which is composed of the latter.

Inasmuch as *each* of these vital units of different orders is distinguished by its individual functions, whenever such a unit coöperates with an "external" factor, we are often interested only in the behavior of the unit and we call this its *reaction*. In a *complete* estimate of the phenomena we should, of course, have to take cognizance of the way in which the external, or more correctly speaking, "other" factor is affected, especially when this happens to be also a living part.

Thus we speak of the *formative reactions* of cells, tissues, organs, or of the whole organism which these go to make up, *e.g.*, of the influence of increased functional stimuli on bones through the activity of the muscles, etc.

Besides the *modi operandi* or energies of *development*, the *modi operandi* or energies of the *maintenance* and of the *involution* of organic forms and their bearers must be investigated by themselves, although it is probable that *maintenance often represents merely the equilibration of diverse components which are also active and formative during development;* and that during subsequent *involution* this *equilibrium is upset by altering, destructive components*. Besides searching for such conditions we must, on the other hand, seek to determine whether *each of these phases has not formative modi operandi peculiar to itself*.

Furthermore, in accordance with the double course of development, viz., the *phyletic* and *ontogenetic*, developmental mechanics must look for the causes, or *modi operandi*, of each of these two courses ; hence an *ontogenetic* and a *phylogenetic developmental mechanics* are to be perfected.

Since the object of the developmental mechanics of *ontogeny* is the investigation of phenomena which are hurried through rapidly in present time, it will, of course, yield greater results than phylogenetic developmental mechanics, the phenomena of which belong in great measure to the past, and, so far as they occur at present, must be carried on with extreme slowness. But in consequence of the intimate causal connections existing between the two, many of the conclusions drawn from the investigation of ontogeny will also throw light on phylogenetic processes ; moreover, phylogeny, even within the limits of its present occurrence, is not entirely inaccessible to investigation ; many a causal connection may be ascertained by means of *experiment*, as has already been shown in the case of artificial selection.

The components with which the doctrine of phylogenesis has hitherto exclusively dealt, viz., *variation* (adaptation) and *heredity*, are still more complicated than the above-mentioned complex components. Nevertheless, this distinction at the same time represents the reduction of extremely diverse phenomena to two, albeit in their special *modi operandi* exceedingly variable, and hence not "constant" or "uniform," components. The word "variation" is to a much greater degree even than the word "heredity" a *collective term* for *results* which are in a certain sense uniform, but which may depend on very diverse *modi operandi*. Hence developmental mechanics has before it the further task of searching out, first, the various constant sub-components of the effects so named, and, second, the causes of these effects.

In this direction, too, encouraging attempts have been made. While Darwin's *doctrine of natural selection* represents only *collective causes* (Aufspeicherungsursachen) of *given* characters on the basis of the survival, — non-extermination, — of the fittest, the new *doctrine of mechanomorphoses* of Julius v. Sachs (8) is already giving us an insight into actually operant, and hence immediate formative causes, — into the formative *modi operandi* of the prehistoric life of organisms.

II. *Methods of Investigation in Developmental Mechanics.*

The causal method of investigation, κατ' ἐξοχὴν, is *experiment*. This statement holds good of the mechanics of development more than of any other line of causal investigation, as will be apparent from the following considerations:

The formative operations occurring in the organism are hidden from sight; we cannot see the ganglion cells of the anterior cornua influencing the development of the muscles, nor increased activity stimulating the growth of organs, nor the substances secreted by cells exerting a chemiotropic attraction on other cells ; indeed, it is not even possible to observe directly that pressure is exerted by cells during growth, nor the passive alterations in the form of parts on which such pressure is exerted. All these operations can only be inferred.

The ascertainment of these operant conditions is, moreover, made still more difficult because the really formative *activity* is carried on so rapidly, as compared with any visible changes, that even in the production of considerable transformations the efficient causes, the *antecedent is*, according to His (9), *almost always in advance of the effect*, or consequence, *by a differential;* even in eventually resulting passive deformations the nature of the processes cannot be ascertained by removing the pressing parts because the form resulting from the pressure has in every case already settled into *internal equilibrium* and lacks only a minimum of adaptation ; for after the removal of the pressing parts a passively deformed structure does not return to its original form, as does a bent rubber tube after the cessation of the bending forces.

Since, moreover, during the normal development of an individual there are always *many changes taking place simultaneously*, we can only conclude from observation of these changes that the *ensemble* of former changes is or may be the cause of the changes which follow ; *but we are not in a position to conclude on what preceding change each single ultimate change depends*.

In accordance with the aphorism: two phenomena which always occur together are causally connected, we can, it is true, deduce *from comparative observations on normal phenom-ena*, without recourse to experiment, many *modi operandi* which obtain among the parts; and these *modi operandi* will have the greater probability the greater and *more varied* the materials of observation.

In this way Balfour (10) deduced the fact that the eggs of sharks, bony fishes, and birds undergo only a partial segmenta-tion from the inhibitory effects on division of the great amount of yolk accumulated in the eggs of these animals.

Nevertheless such conclusions never yield "complete" certainty because the observed connection of the phenomena need not be a direct one, but may depend on the effects of a third unknown component, or components. For the organic processes of the *typical* or normal development of organisms are so incompre-hensibly manifold and enigmatical that, particularly in the beginnings of *exact causal investigations*, we can never deny with assurance the existence of such a common third compo-nent or other components; and the less because in every case only a small part of the secondary or tertiary phenomena fall within the limits of our observation, while all the primary phe-nomena of organic formation are concealed from our view. *Hence, modi operandi may be "ascertained" by means of com-parative observations on normal phenomena, but they cannot be "proved."*

This must always be borne in mind; we can never regard such effects as are concluded from mere observation of typical normal phenomena as perfectly certain; *we must endeavor to obtain direct proofs of these effects.*

It has been shown that in the early cleavage of many eggs the directions of the division-planes follow one another in defi-nite sequence. Bearing in mind that the nuclear spindle lies normally at right angles to the division-plane of the cell, the common result of the directions of these first divisions with ref-erence to the *shape* of the cells in corresponding periods of time, is this: in these first divisions the nuclear spindles place them-selves in the *longest* axis which can be drawn through the cen-ter of mass of the protoplasm. Starting with this statement, it is possible to conclude deductively as to the sequence of the first planes of cleavage. At the same time, this statement is *not certain* so long as it cannot be proved *directly;* for the same typical sequences of division in normal development might be brought about by other, albeit, perhaps, *much more complicated,* but, nevertheless, *typical* operations. More than by a hundred further agreements with the rule in *normal* phenomena, the approximate truth of the above statement was proved by a single experiment, in which by pressing the eggs till they assumed an abnormal form, the sequence of the planes during the early cleavage departed from the normal, but even in this condition the nuclear spindle came to lie in the above-mentioned greater axis. At the same time it was shown, however, that in rare cases the nuclear spindle places itself in the *smallest* axis which can be drawn through the center of mass of the protoplasm, a fact which points at the same time to the operation of *several* factors in the determination of this directive influence (5).

Among biologists there is a tendency derived from the inorganic sciences, *to regard the hypothetical deductions which appear to us to be the "simplest," as having the greatest probability for the very reason that they seem so simple.*

Although much has been done on this assumption, and unfortunately must be done, and although much that is true has already been brought to light, nevertheless this method must always be applied with great reserve to normal biological phenomena, for deeper knowledge shows us that *we have not yet a sufficient insight into the actual mechanisms of development to venture an opinion as to what may be easiest and simplest for these same mechanisms.*

Thus we suppose that we are really simplifying matters when, *e.g.,* we attribute in consequence of functional adaptation many typical and purposive forms to the self-constructive effects of use. The correctness of this *principle* and of its application in many cases has long been capable of direct proof. Nevertheless, we observe that many structures which might be the result of this principle, *e.g.,* the form of joints, the functional structure of the gut, are already established before there is an opportunity for them to exercise their definitive functions. Hence some other mode must operate in producing these structures, a mode, which, it would seem, depends rather on *independently* inherited and *typically* formative forces in individual organs. This latter formative mode appears to us more difficult than the former for the very reason that it requires a whole series of *independent, typically localized individual formations* for the building up of a single structure. But that *this formative mode must in reality be carried out with very great ease* is shown by the difference in the rich and beautiful pattern of birds' plumage in closely allied species, although in every such plumage every feather, characterized as it is by its position on the body and its relation to the other feathers, must have its own typical pattern, differing from that of neighboring feathers in a typical manner.

" *Certainty* " in causal deduction *can only come from experiment*, either from " *artificial* " or from " *nature's* " *experiment*, such as *variation, monstrosity*, or other *pathological phenomena ;* this certainty, however, is only to be obtained by adhering to various precautions which are often difficult to follow.

In an experiment performed under the most favorable conditions, only *one* of the components known to us is or will be changed, and through the results of this change we apprehend those phenomena which are connected with this component.

In practice, however, matters are not so simple ; for in organic objects even after artificial, analytical experiment we often experience the greatest difficulty in tracing back the effects to their *true* causes ; in the first place we are obliged to repeat the experiment often in order to obtain *constant results* and then it must be *modified in various ways* in order that we may be able to determine the true causes. This is because the conditions are so complicated that we do not know the primarily altered components even by means of artificial interference, since, when we suppose we have succeeded in changing only a single component, *accidental* external or internal conditions or unintentional *collateral effects* of our own interference have already affected several components. Only when we are perfectly sure that in reality no other than the single component which we intended to change is affected, are we in a position to draw a definite causal conclusion from a *single* experiment.

This conviction or insight will only rarely be obtained from experiments on organisms. Hence it so often happens that when we believe we have experimented under the very same conditions and in the same manner as on a former occasion, we nevertheless obtain different results. So long as we do not arrive at the same result, at least after several repetitions of the same experiment, we must not permit ourselves to draw any conclusion whatsoever. And now that we are in the first stages of our investigations, without having any survey of the *modi operandi* which may occur, it will often be necessary *to use as many methods as possible in experimenting on the same subject;* and only when these different experiments point to the same causal connection should we assume that this is the true one.

With the aid of such experiments we are in a position on the one hand to test the relationships which are *determined* by comparative study of the normal forms, and on the other hand to obtain — yes, to extort — an answer to newly arising questions.

Before we can establish the *causal modi operandi according to their qualities,* we must *first determine the parts between which formative operations take place,* i.e., we must determine the " *locality* " of the formative operations. With reference to the *single circumscribed structure or part,* this means that we must ascertain whether the causes of its formation lie within itself or whether external influences are necessary to its formation.

The rôle which the different causes that take part in a formation play in its production may be a very unequal one.

Inasmuch as some singular notions and terminology have gone abroad concerning *causes of different dignity,* it seems proper in this place to go somewhat into details for the sake of paving the way towards greater uniformity of opinion.

All the components whose *temporary* and *local coincidence* is necessary to produce a certain effect, constitute in their totality the "*whole cause*" of the effect. Of these components we often call those with the commencement of which the effect *begins* (*i.e.*, the last preceding *event*), the *cause* of the effect, while the components which were previously and continuously present (*i.e.*, the *permanent facts*), are known as the *preëxisting conditions*. This is, however, an arbitrary distinction, and one which is detrimental to our quest for *complete* knowledge. The essential point is this: *All the components* of an effect must exist *beforehand*, but they need not all "*begin*" immediately before.

It seemed to me useful, in order to further the special aims which we have in view, to introduce a different distinction of cause and preëxisting condition, although this distinction, too, is somewhat arbitrary.

I have called such components "CAUSES," or, better, "SPECIFIC CAUSES," "SPECIFIC COMPONENTS" of a *process of organic formation*, as condition the "*specific nature*" *of the process*, while the other components which are equally essential to the starting in of the phenomena, but which, like heat and oxygen, do not determine the character of the formation, were called "PREËXISTING CONDITIONS," "INDIFFERENT CAUSES," or "INDIFFERENT COMPONENTS" (11).

If our endeavors be directed not to the *qualitative* cause of the phenomenon but only to the cause of the *place*, *time*, or *magnitude* of the same, we must designate as "*specific*" *causes of these circumstances* those causes which condition the given circumstance.

The theory of this unequal participation of the components in conditioning the *specific nature of the resultant* requires further elaboration.

Starting with the view of the different functions of the components of the same process and consequently with a preference for the components which condition the specific nature of the phenomenon investigated, I have designated as "SELF-DIFFERENTIATION" *of the circumscribed or presumably circumscribed structure or part*, that change, whose specific causes (in the sense just defined) lie within the formal structure or part itself; and this expression would be employed even when the admission of energy from without, in the form of heat, oxygen, etc., is necessary. In order to distinguish the two cases the term " COMPLETE SELF-DIFFERENTIATION " was employed when *all* the components lie within the formed part itself, while " INCOMPLETE SELF-DIFFERENTIATION " obtains when the accession of energy from without is required, in so far as this energy represents only the preëxisting condition of the formative operation in the sense above accepted; but since, nevertheless, the accession of energy from without in the form of heat, light, gaseous, and liquid nutriment, is in varying quantity necessary to the development of the eggs of different animals, but does not determine whether an egg is to develop into a chick, a frog, or a fish, or whether the lung is to be laid down at a particular spot in the embryo, *the development of the egg would be more accurately designated as " incomplete self-differentiation."*

As was set forth above, self-differentiation in the strict dynamical and analytical sense, can, of course, have no existence, since every change in a phenomenon must depend on reciprocal operations. Since the concept " self-differentiation " is, accordingly, not processual but merely topographical, implying something with regard to the *locality* of the causes of the formative process, whenever it is employed, the particular circumscribed structure to which it refers must be mentioned.

" DEPENDENT DIFFERENTIATION " is a change in which one or more of the components that condition the specific formation, operate from without on the circumscribed or presumably circumscribed part to be formed ; and " PASSIVE DIFFERENTIATION " occurs when all of the components of the respective formative process of a given part operate from without, as, *e.g.*, in the modeling of a figure in clay or wax.

Self-differentiation and dependent differentiation may occur in the most varied combination either simultaneously or successively.

Thus the normal formation of skeletal structures like the tibia is very probably partially due to self-differentiation, because, presumably apart from external influences, there arises from the given Anlage-material a rather long skeletal structure with a thickening at its proximal end; but in other respects this formation is due to dependent differentiation, since the finer details of structure, like the surfaces of the joints and the three-sided shape of the diaphysis, are conditioned by the operations of neighboring parts.

The segmentation of the common arborescent glands into lobules appears to be conditioned by the formative operations of the epithelia and hence of the specific parts, and, so far as this is true, the segmentation is a *self-differentiation of the glandular substance*. In the liver, however, which is a reticular gland, the normal size and form of the lobules and also the lobular segmentation itself appears to be conditioned by the blood-vessels — on the one hand by the requisite length of the capillaries, and on the other by the peculiarity in the ramification of the portal vein, which during its growth develops *dichotomic* branches in its capillary network. Hence *the acinous segmentation of the liver parenchyma represents a differentiation of the glandular substance depending on the vascular system.*

After, or at the same time as the actual ascertainment of such "*local*" *conditions of the formative causes*, we shall endeavor to look for factors which condition the *magnitude* and *direction* of the formative processes ; simultaneously, or even before this, we may be able to ascertain also the *time* when many of these formations are reduced to a *norm*, as, *e.g.*, the *time* when the *direction* of the median sagittal plane of the embryo is determined ; for it is not necessary that these formative conditions be first conditioned when the ultimate forms first become visible.

On the contrary, in the perfectly normal, *i.e., perfectly typical,* course of the individual development, all the typical structures must at the very latest be in some way conditioned in the fertilized egg, either *implicite* in their earliest components, or *explicite* in already visible Anlagen. Nevertheless, we must assume that there is really no such thing as *perfectly* typical development (12), but that in every individual development greater or less disturbances take place, which are compensated by the putting into action of regulating mechanisms. Accurately speaking, therefore, we should only have to determine in respect of time, *within what preceding developmental phases structures which are not visible till sometime afterward can no longer be varied by disturbing influences;* and in respect of form, *what* preceding visible or invisible structures condition every formation that is later observable, as, *e.g.*, in the case of the median sagittal plane of the embryo which is *normally* conditioned by the first cleavage plane, and this in turn by the axis of the copulation of the male and female pronuclei.

Ultimately we shall attempt to get at the *causal modi operandi*, by attempting to ascertain their *quality*, and to trace out the more general *modi operandi*, of the combination of which a given effect is itself only a special case.

For all this *analytical* experiment gives us ample opportunity. By isolating, transposing, destroying, weakening, stimulating, false union, passive deformation, changing the diet and the functional size of the parts of eggs, embryos, or more developed organisms, by the application of unaccustomed agencies like light, heat, electricity, and by the withdrawal of customary influences, we may be able *to ascertain a great many formative operations in the parts of organisms.* Thus we may, perhaps, determine the possible influence of the muscles in the formation of the joints and sockets by cutting the sinews of the biceps and triceps brachii in very young animals and sewing them on again with transposed insertions; by cutting out transverse wedge-shaped pieces from the longer bones and feeding with madder, it may be possible to learn something of the processes of functional adaptation in the structure of the bones and hence of their immediate relations.

By such artificial interference we shall in the first place be able to establish the occurrence of dependent differentiation and hence of differentiating reciprocal effects in *such* parts as are far enough removed from one another to be isolated by the crude means at our disposal, without their vitality being destroyed by the harmful vicinity of the wounded region.

Even now several results seem to show that during the course of *normal* development, the "specific causes" of many differentiations lie almost entirely within the altered parts, even in very small parts, so that, therefore, areas of independent differentiation may at an early stage comprise a single or only a few cells.   The investigation of such narrowly localized processes of differentiation is attended with much greater difficulties ; and since, moreover, *the fundamental formative processes*, viz., assimilation, growth, self-movement, and the qualitative differentiation of cells take place altogether or, at least, in the first instance within the province of the invisibly minute, it will be necessary, in order to clear up these fundamental processes, *to make as much or even more use of hypotheses, as physicists and chemists are compelled to do* when they cope with the fundamental processes of their respective sciences.   And just as in these sciences, we shall have to regard those assumptions as approximating most nearly to the truth which explain the most facts and permit of the successful prediction of new facts ; and *ceteris paribus* we shall prefer that explanation which appears to be the "simplest," not forgetting, however, that we may easily fall into error on this point for the reasons above set forth.

Experiment on living beings is quite peculiar and apt to be misleading, in that in many cases, like *mutilations* and certain *disturbances of the arrangement* of parts with respect to one another, conditions arise in which the organism does not react with the formative mechanisms of *direct or normal development*, but with the regulative and regenerative mechanisms of *indirect development, or regeneration* (13).

Indirect development runs its course in great measure under the *regulating reciprocal activities of many*, or, as in the case of great defects and disturbances in lower animals, for a time at least, of *all* parts of the organism ; it differs essentially in this respect from the *direct or typical development* of the fertilized ovum, which goes on in the absence of any interference, or even for a short time after the cessation of the interference, and often completes its course with extreme *self-differentiation* of circumscribed parts.   (*Within* these, of course, the changes depend on the *reciprocal operations of the parts*.)

The modi operandi of each of these two varieties of development must be investigated.

In the setting to work of the mechanisms of *indirect* development lies, however, *one of the greatest hindrances to the investigation of normal formative modes* of direct development.

In those low organisms in which regeneration steps in promptly after *mutilation* or after *disturbance in the arrangement* of parts, the value of the experiment is much lessened when it is intended for the investigation of the *normal* methods of development. On the other hand, the higher organisms are more advantageous in that their regulatory mechanisms, especially during the later stages of development, are much weaker in their manifestation and in part much more difficult to call into activity, *i.e.*, they set in much later after the disturbing influence than they do in lower animals.

This favorable circumstance enables us to investigate exhaustibly by means of experiment the processes of normal development in the organisms which rank next to ourselves.

Owing to the fact that these two typically different kinds of development, as well as the rôle they play in the reactions of animals subjected to experiment, have not been kept distinct heretofore by most experimenters, recent experimental investigation has been productive of more confusion than enlightenment; quite apart from the fact that the observations themselves leave much to be desired in point of accuracy and completeness, perhaps for the reason that we do not yet sufficiently appreciate how much more expenditure of patient observation is required in experimental investigation than in current descriptive embryological investigation. In the latter we are already sufficiently advanced to be able to recognize the different developmental stages, and we often know when the stage of immediate interest will make its appearance; whereas unusual experimental interference may *at any time bring forth something new*, so that in order to follow up the subject it is often necessary to observe continuously, or at least frequently, by day and night.

We must not conceal from ourselves the fact that the causal investigation of organisms is one of the most difficult, if not the most difficult, problem which the human intellect has attempted to solve, and that this investigation, like every causal science, can never reach completeness, since every new cause ascertained only gives rise to fresh questions regarding the cause of this cause.

Inasmuch as many of its problems are nearly or quite insoluble by means of experimental investigation, *developmental mechanics must needs, so far as possible, seek to utilize for its own ends, all the kinds and ways of causal investigation of organisms and the results thereby attained,* and not cast aside as useless any biological discipline in silly conceit. Developmental mechanics should, moreover, cultivate the analysis of formative processes into constant "complex components" to a greater extent, if anything, than the ascertainment of simple components.

This conception of the methods of investigation *first* to be undertaken in developmental mechanics differs essentially from the views of many contemporaneous workers in the same field, who believe that descriptive and comparative anatomy as well as embryological investigation are of little value to developmental mechanics. This opinion is held by authors who see the *present* task of developmental mechanics in the immediate reduction of organic formative processes to purely inorganic, physico-chemical components (14).

If, however, we limit ourselves to that which is *possible* at present, we can regard *this* task only as a *final goal*, which for the present, and even for some time to come, we shall approach in a *direct* path only at a relatively slow pace ; still we are not to cease in our endeavors "to reduce the formative forces of the animal body to the general forces or vital tendencies of the world as a whole," as K. E. von Baer has said (15).

It is evidently advantageous, and will be productive of much important information, if we endeavor to reproduce *synthetically* in an inorganic way *structures, forms, and processes which resemble as closely as possible,* or are the same as those of the organic world. This has been done by G. Berthold, Errera, and more recently, and with marked success, by O. Bütschli.

Were we, however, to follow this as the *only* method of procedure, and, in accordance therewith, to attempt the investigation only of those processes which resolve themselves at once into simple components, or from which at least such components may at once be *split off*, we should very soon reach a limit at which we should be brought to a standstill ; for the majority of organic processes are far too complicated in their conditioning to admit of immediate reduction to physico-chemical *modi operandi*. And even in cases where it is claimed that such a reduction has been brought about, it appears that the part which the simple components contribute to the formation in question, as compared with that of the coöperant complex components, has been considerably overestimated.

If we would advance without interruption, we shall have to be content for many years to come with an analysis into complex components.

While thus in some quarters the possibility of a physical explanation, so far as it is attainable at present, is considerably overestimated, it appears that in another quarter our possible attainments in this direction are, on the whole, essentially underestimated, so that organic structure is claimed to be incapable of any explanation, and only to be deduced teleologically.

We may be easily misled to such a metaphysical conclusion by the facts of regeneration, and also by the observations recently made by Driesch on the origin of normally formed products after *extreme* interference during the early development, viz., during the cleavage stages. Although these processes actually do produce the impression that mechanical operations are inadequate, and that the purpose of bringing about the typical form as a whole must step in actively, still we are bound not to entertain such a supposition, at least with our present limited insight, *till every other possibility has been with certainty excluded.* This is at present by no means the case. For in regeneration there is still extant a portion of the typical whole, a portion, moreover, in which the *whole itself* may be supposed to be contained *implicite* in the form of germ-plasm, and hence in an undeveloped condition ; this regeneration-plasma being called into activity, may, thereupon, restore the whole *explicite*. *From this source* is brought forth again the *typical* form, after its kind, and, what is worthy of special consideration, often in a somewhat *defective* manner. The problem is not, therefore, one of a peculiar nature, nor one which involves a leading principle, but refers solely to the *special process whereby* the normal form is restored. The same holds good also with respect to the manifestations of the postulated regenerative-plasma in cases where development is disturbed during the cleavage stages.

The continuity of typical formation, the continuity of the typically developed and undeveloped material of formation is, therefore, not interrupted by these irregular processes, and, no matter how difficult it may be to form a conception of the details of the phenomena, *there is still no urgent reason for assuming a metaphysical process.*

"*Incidit in Scyllam, qui vult vitare Charybdim*" is particularly applicable to the investigator in the field of developmental mechanics. The *too simply mechanical* and the *metaphysical conception* represent the Scylla and the Charybdis, to steer one's course between which is indeed a difficult task, a task which few have hitherto accomplished. It cannot, however, be denied that the seductiveness of the latter views has been increasing with the increase in our knowledge.

The *least productive method* of carrying on developmental mechanics is to start out in the very beginnings of exact investigation from the limited number of facts at our disposal, and to pour forth numerous and long-winded essays on the length to which our understanding can go in this field and on the rôles which opposing formative principles play during developmental processes.

It is true that in order to understand the problems before us it was necessary to elucidate more clearly the old contrasts between Evolution and Epigenesis, but this was not for the purpose of producing endless theoretical disquisitions, but *with the aim of establishing a basis for exact investigation* (16). Still we must regard as useful the attempt to bring together all the facts which were supposed to support each of the possible views. Continued discussion, however, and the premature expression and maintenance of final one-sided opinions on these still unknown conditions, can only injure the reputation of our immature investigation along causal lines, and withdraw the few who have devoted themselves to the subject from more productive activity.

III. *The Relations of Developmental Mechanics to the Other Biological Disciplines.*

The branches of Biology hitherto recognized, viz., descriptive *zoölogy*, *anatomy*, *embryology*, and *physiology*, represent the essential prerequisites of developmental mechanics, for it is they that teach us the facts in forms and processes, the causal explanation of the latter being the province of the discipline we are discussing.

Because they depend on comparison of structure, anatomy and embryology are also productive of causal information to the extent that such comparison can take the place of experiment.

This substitution cannot be a complete one for the logical reasons presented above. Nevertheless, *comparative anatomy and comparative embryology are the means of ascertaining many causal relations* between the parts of organisms, and these relations, in so far as they rest on a sufficient mass of observations, lack only the direct proof of artificial or natural experiment to become certainty. *In so far as these disciplines reveal causal information, they are themselves developmental mechanics,* and inasmuch as they do and have done this to a very great extent, they represent disciplines which are only historically separated from developmental mechanics.

The new character which these causal investigations have acquired in recent times, and will continue to acquire, is the use of *analytical* experiment, together with the endeavor to collect together all causal information, and to raise causal investigation to the dignity of a principal aim, — an aim in itself.

Thus phylogenetic and ontogenetic developmental mechanics receive from the older branches of biology besides their problems much causal information, and still more guidance to such information. *The methods with which this knowledge has been acquired will continue to be necessary to developmental mechanics even in future,* since many causal problems are scarcely accessible to experimental investigation, and since, moreover, the correct interpretation of the results of experiment is often fraught with such difficulty, that every possible aid from other sources must be utilized.

Still developmental mechanics will be of more or less service to these morphological disciplines in return for what it is continually receiving from them.

It will open the eyes of the descriptive observer to many structural relations hitherto overlooked ; structures which have been scarcely appreciated will acquire a deeper significance ; many a problem arising from descriptive study and incapable of solution through observation on normal phenomena will be elucidated, and the causal deductions of these sciences will be corrected or established on a firmer basis. Thus the doctrine of the *transposition of cells* during embryonic formation — a doctrine which has of late been greatly expanded by His — will be *proved* to be correct only by experiment, and tested as to the extent to which it is claimed to obtain, and traced back to its causes. In like manner our ideas derived from comparison of different *phases of cell-division* require direct experimental proof, or confirmation and extension with respect to the immediate causal interrelations of these processes. It was only through causal observation that life was infused into the dead facts of corrosion anatomy, when the *laws which govern the ramification of blood-vessels* were discovered.

Comparative anatomy will be able to receive a great deal of assistance from developmental mechanics, especially *in extending the problems with which it deals*. As comparative anatomy endeavors to ascertain the genetic connection, the "Stammbaum" of organisms, it is itself essentially a causal science. It analyzes structures into the two components, *variation* and *heredity*. It is true that both of these, as understood in comparative anatomy, are general formative principles, but, in the first place, they are of much greater diversity than the complex components given above as illustrations, and in the second place, they are not uniform, *i.e.*, not always constant in their modes of operating.

Heredity is a constant principle, always *operating* in the same way, only in so far as it depends, according to Weismann and others, on the continuity and variations of the *germ-plasm*, and hence on *assimilation*. When we are dealing besides with the inheritance of somatogenic, or so-called acquired characters, the same word is used to designate *modi operandi* of a totally different nature.

The concept *variation* (adaptation) comprises so many different operations that Haeckel has established for them a whole series of "laws" (19). Both heredity and variation, however, are in urgent need of causal explanation, *i.e.*, of analysis into their uniformly operant components. This analysis is one of the tasks of developmental mechanics. This is true also of *cœnogenesis* and of the so-called *"fundamental law of biogenesis."*

The hypotheses which *comparative anatomy*, like every other science, continually employs, *have essentially the character of developmental mechanics.*

As this fact does not seem to be sufficiently well known, a few illustrations may be adduced here.

Gegenbaur rejects the homology of the ventral nerve-cord with the spinal cord (20) mainly for the reason that he regards the *difference in the respective "positions"* of the two organs as much more important than the *agreement* of their occurrence throughout the whole length of the animal, their metameric segmentation, similarity of ramification, and composition of the same form-elements. This opinion rests upon the assumption that in phylogeny an organ may more easily arise *anew* and independently of a preëxisting organ with which it has in common the same biological constituents, essentially the same distribution, the same segmentation, and the same function, than that the latter organ should have changed its *position* to such an extent, viz., from the ventral to the dorsal side of the animal.[1]

As will be seen, this assumption is purely one of developmental mechanics and was certainly a bold hypothesis in the state of developmental mechanics at that time; and although we do not doubt its truth in this particular case, Gegenbaur

---

[1] I have taken the liberty of correcting an obvious *lapsus calami* in this sentence. — W. M. W.

himself would hardly regard it as true *in general,* but only in respect of such axial organs as would have to shift their position through an angle of 180° to the opposite side of the body.

The morphological inequality of the upper lobes of both human lungs assumed by Aeby (21) — an inequality which he deduces from the fact that the bronchus from the right side takes an eparterial, that of the left, a hyparterial course, so that the left lung lacks an equivalent of the right upper lobe — also rests upon the developmental mechanical assumption that the relations of *position* of the air-passage to the blood-passage are essentially more constant, *i.e.,* may vary with less facility than the *shape* of the portions of the lung to which these two passages lead.   This assumption, though doubtful, is supported by the fact — also of a developmental mechanical nature — that the lung has little shape of its own, but adapts its form largely to its environment.

The fundamental law propounded by Wiedersheim (22) as the result of extensive comparative investigation, "that the impulse to the development of the appendicular skeleton in vertebrates always starts from the periphery, and that the central (girdle) portions are only secondarily developed under the formative influence of the free appendages," is, as will be seen, also of a purely developmental mechanical nature, and requires further developmental mechanical substantiation and analysis.   This is also the case with the important conception of imitative homology, or parhomology, introduced by Fürbringer (23).

Although these examples have been adduced without special selection, they nevertheless show clearly how comparative anatomy is continually assigning problems to developmental mechanics by making that science acquainted with new operations, and how, on the other hand, developmental mechanics, by devoting itself to the solution of these problems, is becoming the continuation and at the same time the mainstay of comparative anatomy.

As long as comparative anatomy attempted to establish only the *main course* of development in the animal kingdom, following in a general way the continuous development of forms only through the *classes* of each type, comparison of different forms showed that essentially and unequivocally the same course of progressive development is followed by nearly all systems of organs. But in further approximations of a higher degree, viz., in tracing that development through the orders, families, genera, and species, even to the individual, so many incongruities in the development of organ systems and organs made their appearance, that comparative anatomy has been compelled to call in the assistance of quite a number of developmental mechanical hypotheses, for the correctness of which only experimental tests can give complete security.

Even the *appreciation of " essential" or " unessential" agreements or differences*, an appreciation which is continually necessary in the phylogenetic explanation of comparative observations on form, *in ultimate analysis always shows itself to be of a developmental mechanical character.*

Since developmental mechanics, perhaps for some time to come, or at least in the beginning, will pursue its own course, it would be encouraging if comparative anatomists would themselves resort to experimentation for the purpose of solving, so far as possible in a short time, the problems in which they are interested, *e.g.*, the continually recurring main question, as to what are actually — not in a formal, but in a developmental sense — " *slight* " or " *easy* " *variations ;* whether *the number* of organs may be *increased* " easily " (*i.e.*, by a simple interference and hence by a correspondingly slight accident), as perhaps by the passive infolding of a somite, by the splitting of a shoot, or by linear pressure on the same in a direction contrary to its direction of growth, and further, in case these attempts are successful, whether or not such newly formed organs at once attain to the full differentiation of the former ones ; further, whether, inversely, a *decrease in the number* of organs may be " easily " brought about, perhaps by inhibiting a normal infolding or constriction or by compression and resulting concrescence ; whether in these cases according to the earlier or later stage of development, during which such interference is applied, the united parts may at once become perfectly simple or still retain traces of their double origin, etc.

Of course these would not be *hereditary* changes; on which account, the essential results of these experiments could only be utilized in explaining *individual* variations with reference to their representing "reversions" or "monstrosities." Hence it would be of greater importance to ascertain to what extent after artificial *local* changes in an embryo, changes make their appearance in other organs — no matter whether these bear functional correlations to the affected regions or not — since in the case of the same primary or inherited change the secondary changes would then also be "inherited." Moreover, by raising animals that are born without fore limbs or have been deprived of them, it may be possible to ascertain to what extent such animals, being compelled from the first to adopt a method of locomotion, like jumping, which is foreign to their species, are nevertheless able by direct adaptation to this mode of progression, to develop the requisite proportions in the length of the skeletal parts and in the size of the lever-arms of the muscles, and whether in these respects Lamarck's theory is confirmed or refuted.

In the introduction to his *Morphologisches Jahrbuch* Carl Gegenbaur gave expression to the following words full of insight : "Indeed the time will come when morphology, too, will be conscious of the mutability of its aims and aspirations and when other problems and methods will take the place of those with which we busy ourselves at present." This new end is that of developmental mechanics — the investigation of the causes of the forms of organisms.

But it will be a long time before it takes the place of "the aim" of morphology. In the sense of the comparative anatomist, this can only come about when this science has reached the measure of its possible perfection. In the last instance both tendencies have the same aim and it is through *coöperation* that an approach to this aim will be most facilitated.

We must also define our position with respect to *Physiology*. This science in its fullest sense embraces *all the functions of life*. Developmental mechanics represents an integral part of this science, and after it has reached its development it will be the largest and most essential part. But alongside of human and animal physiology as it is almost exclusively carried on by its representatives at present, under the stress of immediate questions; alongside of this science which treats of the maintenance-function of parts already established, usually to the exclusion of the formative functions of maintenance; alongside of the residual "*science of the mere keeping a-going of the living machine*," whereby the functions most difficult of comprehension, viz., those of the construction, formation, and the maintenance of that which is formed, remain unheeded and uninvestigated — the science of the causes of this formative activity constitutes an essentially independent branch.

Since, however, the performance of a function, even in already developed organs, has a *formative effect* in consequence of "*functional adaptation*" to magnitudes of function which have been increased for a considerable time beyond the common mean, or depressed below it, this *doctrine of mere machine-activity* is of importance to developmental mechanics, and many of the results of its investigation may be of service to the latter, so that we must also remain in close touch with this kind of physiology. But quite as great or even greater will be the assistance which later on this physiology will receive from an insight into the causes of the formation and maintenance of structure.

Since in plant life the *formative functions* greatly predominate over the *functions of maintenance* (Betriebsfunctionen), owing to the absence of the nervous and muscular systems and sense organs, and since, moreover, plants are more easily accessible to experiment than animal organisms, *plant physiology* has been spared the onesidedness which exists in animal physiology; thanks to the investigations of such men as Julius v. Sachs, Wiesner, Pfeffer, Strasburger, Berthold, de Vries, Voechting, Klebs, and others, it has already become in a great measure developmental mechanics in the full sense of the word, and has far outstripped the developmental mechanics of animal organisms.

The causal tendency of Phytomorphology was considerably advanced by the fact that *plant forms*, being fixed to a particular spot and hence much more exposed to external influences and to these in part in constant directions, *are influenced even in their typical morphology to a great extent by " external" factors, whereas the " typical" structure of animals, which are capable of active locomotion, is in great measure independent of external formative influences* and consists, apart from certain functions of superficial parts, in self-differentiation. It is, however, much more difficult to understand the *internal* than the external factors and the reactions to the same.

In *sessile* animals J. Loeb (24) has recently discovered differentiating effects of gravity on the organism, like those observed in plants. For example an inverted piece of a hydroid polyp will produce *roots* at its *lower* and *shoots* at its *upper* end. But we must be careful not to extend this occurrence to other animals, as has already been done, thus ignoring the causal implication in the sessile mode of life, and ascribing in all animals a differentiating effect to gravity, especially when irreproachable experiments have already proved the opposite in the case of other animals.

Of particularly great importance to developmental mechanics, are, furthermore, many of the results of the PATHOLOGICAL SCIENCES.

Looking aside from the cases in which *immediate death* is brought about by a sudden stopping or disturbance of the functions which are necessary to keep the machine going, we observe in every *primary disturbance*, no matter how it may be caused, *secondary changes* intervening, which even though they be *merely functional* at first, nevertheless gradually lead to *formative* changes.

In this manner these secondary formative changes give us evidence of formative interrelations, formative modi operandi of parts one upon another, an understanding of which is essential to our purpose.

But even here, as in the effect of an experiment, we must first ascertain whether these pathologically formative *modi operandi* enable us to draw any conclusions whatever with respect to *normal* operations, or whether under *abnormal* conditions abnormal modes of reaction may also occur, and hence processes which do not occur at all in normal phenomena.

To sum up the results of observation in the pathology of the *higher vertebrates*, we may say that pathology is essentially the doctrine of phenomena which are in themselves normal, but which manifest themselves in the wrong place, at the wrong time, or in the wrong magnitude or direction; for all pathological *processes*, a few *kinds of decay* (like amyloid and waxy degeneration) excepted, also occur as normal phenomena.

Hence there do not occur in the pathological conditions of these animals any *modi operandi which are foreign to normal development or any new substantive or even productively formative modi operandi;* and hence in case of secondary changes pathology has only to investigate the way in which the organism makes use of its normal modes of formation and reaction during or after disturbances of the normal conditions.

Of course these results of pathology hold good also of *artificial* experiments. *We are able to conclude*, therefore, from the reactions which take place after experimental or pathological changes *as to the modi operandi which also occur under normal conditions*, but which operate normally with different intensity and at a different time.

On the other hand, whenever regeneration of destroyed parts occurs, the mechanisms of *indirect* development are put into activity. These were referred to above.

Here we are concerned with the *secondary changes of other parts*, which following upon primary disturbance are either themselves disturbances; in this case they indicate that the primarily affected part is necessary to the maintenance or development of the secondarily affected part, and hence in some way participates in its production, thus exercising a " trophic " influence upon it.[1]

Such conditions follow from the secondary atrophy of the sensory or motor nuclei of the brain and spinal cord when their respective peripheral end-organs are removed soon after birth, and inversely from the aplasia of the muscles after destruction of the motor ganglion cells of the anterior cornua in infantile paralysis; from the degeneration of the nerves when they are separated from their respective ganglion cells, etc.

[1] There is only one alternative mentioned in this sentence, the other clause having been omitted. — W. M. W.

The following conditions point to *still more enigmatical connections:* the disturbance in the development of the brain in congenital defect of both suprarenals, the origin of cretinism and myxœdema after complete extirpation of the thyroid, the default in development of the secondary sexual characters, such as the female habitus, the female mammæ, the male habitus, the beard, the male voice after extirpation of the sexual glands; other cases are unilateral visual atrophy, symmetrical gangrene of the toes and fingers, etc.

In an extensive series of other cases, primary disturbance or destruction of one part is followed by a *compensatory hypertrophy* of other parts of the same kind, which take on the function of the disturbed parts. On such manifestations of *functional adaptation* mainly depends — regeneration being insignificant in man — the very important principle of the *equalization of disturbances* after pathological changes, a principle which has of late been thoroughly studied in all its bearings by Nothnagel (25).

Of a contrary nature is the enigmatical compensatory hypertrophy of *non-functioning organs,* e.g., of the milk-glands of young animals.

Besides such trophic and functional correlations, many *other formative correlations* make their appearance during pathological processes. A *mechanical equilibrium of parts* under normal conditions is indicated by disturbances like the bending outwards of the teeth when the tongue is abnormally large, the triangular shape assumed by the previously round tibia when the muscles of the leg are developed, and the return to the rounded contour with the atrophy of the muscles in spinal infantile paralysis, the hypertrophy of the interstitial connective tissue following the atrophy of the specific tissues of organs, the proliferation of the pavement epithelium of the outer surface of the body into cavities like those of the nose, mammary glands, ureters, and bladder, which are normally lined with a different epithelium ; or the proliferation of the vaginal epithelium into the uterus.

To the same category belong the *formative reactions to well known external influences, i.e.,* influences coming from without the parts affected ; the formation of bones in connective tissue that has been subjected to mechanical impact (" Reit- " and " Exercierknochen ") ; occasional progressive ossifications like *leontiasis ossea* after a single injury; further, the formation of giant cells around dead or dying parts (around foreign bodies), around bones which are no longer supplied with nutriment, or which have become disarticulated, the formation of blisters under skin which has been subjected to repeated pressure or displacement, the formation of the placenta materna on any part of the peritoneum in extra-uterine pregnancy, the formation of new capillaries from those already existing in consequence of an increased demand for nutrition, even when this demand is occasioned by the presence of a body foreign to the particular region (metastatic tumor), together with an increase in size in the afferent and efferent vessels of the region, etc.

The fact that transplanted pieces of skin, like artificial noses, gradually acquire connections with the sensory pathways, indicates that the sensory nerves continue to send out processes in all directions till every region is supplied from one, or normally from two sensory branches ; this is evidence, at the same time, of a peculiar touch which the parts supplied with sensory nerves keep with one another or with the sensory nerves of neighboring parts.

The ends of broken bones which are not bound together and hence movable on each other, gradually develop a joint with the circumjacent connective tissue. Since the normal joints are laid down and developed without any movement of the kind, this pseudarthrosis corresponds only to the *further development* of an already formed normal joint in adaptation to an individual requirement.

Peculiar properties of life are evinced furthermore by the hypertrophy of connective tissue and young epiphysial cartilage or bone in stoppage-hyperæmia, whereas, in contradistinction to this, the specifically functional portions of glands, muscles, and of the central nervous system, are injured by such hyperæmia ; further, the tendency of like parts to grow together in synophthalmia, etc. Many authors will be inclined to include here the formation and retention of bones in places protected from pressure (in reality only apparently thus protected) like the arachnoidea, dura-mater, in the atrophied eyeball.

The property of self-maintenance or self-differentiation of parts is evinced by the development of very minute detached portions of tumors which may be carried anywhere by the blood current and grow to be secondary tumors of the same morphological character as the primary tumor ; the development of sporadic masses of gray brain-substance ; the retention of the normal structure in abstricted pieces of the retina lying outside the eye ; the formation of hair and teeth in dermoid cysts ; the teratomata ; the healing over of transplanted skin, bones, eyeballs, etc.

To these examples of the *important developmental mechanical results of pathological research* should be added further those cases of aberrations from the normal which accrue from a study of *monsters*, and the lesser deviations designated as *varieties*.

Besides the varieties which may fall under the observation of anatomists, there are a great number of these " *experiments of nature* " to which especially *pathological anatomists* and *clinicians* have access.

It would, therefore, be most serviceable and advantageous to developmental mechanics if those investigators to whom such phenomena present themselves were more mindful than they have been heretofore of *the importance of these facts in ascertaining normal formative causes*, and if they would for this reason endeavor to collect *all the formative modi operandi of which there is evidence, together with more accurate data concerning their magnitude and time relations, their mode of operating, their connections, and remoter causes*.

It is probably best to begin with an attempt *to formulate concisely* every such phenomenon as a *modus operandi*. Such an attempt shows at once the unsatisfactory condition of our present knowledge, and there follows as a matter of course the necessity of rendering this knowledge more complete.

The same purpose would be served by many observations which pathologists might make during experimentation undertaken with other aims in view. Thus, *e.g.*, in experiments on the effects of hunger, protracted fever, chronic poisoning, or of any other chronic disturbance like paralysis, etc., *a useful extermination of cells*, hitherto unnoticed by pathologists, always takes place — an extermination, the magnitude and extent of which depends upon the still unknown magnitude of *qualitative* variations among the like cells of a single organ. Under such circumstances the cells which happen to be least able to resist the noxious influences must *ceteris paribus* be the first to perish, and for this very reason after these cells have been supplanted by the offspring of qualitatively more resistent cells, the whole organism, or in the case of local affections, the organ in question, must have become better able to resist these particular noxious influences. (This does not exclude the possibility that in special cases the resistance may be at the same time diminished by other factors.) *By means of hunger, e.g., the organism is transformed by a process of selection into a saving machine,* because those cells which require much nutriment will be the first to starve. Such an *internal selection* must also occur among the *variations* in nourishment and activity during the course of *normal* vital processes, but to a considerably less extent and in a manner more difficult to determine ; hence we may expect that these conditions will be first elucidated in pathological cases of a grosser character.

Since, moreover, pathologists, representing as they do the science of phenomena which are to a considerable extent normal though occurring under abnormal conditions, take a real interest in learning to comprehend normal modes of formation, it will probably be the case in future more often than at present, that these investigators will experiment with the express purpose of ascertaining and analyzing normal modes of formation. This has already been done with success by surgeons in the case of the *modi operandi* of bone-formation.

The "Archiv für Entwickelungsmechanik" will be glad to welcome every such contribution from clinicians and comparative anatomists.

The advantage that will accrue in the first instance to developmental mechanics from such contributions will revert to the service of the clinical disciplines, when once the *modi operandi* of formation and maintenance and their causal relations shall be to a considerable extent understood. For in this way we shall acquire a deeper insight into pathological changes and at the same time a foundation for a therapeutics scientific in the true sense of the word and based upon adequate understanding.

Just as developmental mechanics utilizes for its own purposes all methods which may be productive of causal understanding and all biological disciplines, so does it embrace as its field of investigation all living things, from the lowest Protista to the highest animal and vegetable organisms.

Accordingly *these Archives will accept causal essays on all biological subjects*, but as it does not propose to compete on their own special grounds with periodicals devoted to special subjects, *only those biological papers will be included which directly pursue a causal aim and for which the material has been collected and elaborated with this end in view.*

Descriptive papers, however, containing only occasional suppositions of a causal nature, or even apodictic assertions without any attempt to support these assumptions by comparison of the different pertinent facts, fall outside the scope of these Archives. But it may be suggested to such authors as desire their causal remarks to be preserved, to send their papers to the editor, with an indication of the passages in question, so that attention may be called to them incidentally, perhaps in the form of an essay.

Papers of a comparative anatomical nature which reduce the forms of organisms exclusively to the factors of variation and heredity, without attempting any *further analysis* of these "inconstant" complex components, also lie outside the territory covered by our Archives, since such preliminary analysis together with the ascertainment of descent, properly belongs to the field of comparative anatomy.

It is much to be wished that in concluding *every contribution* which appears in these Archives, *the causal results be concisely summarized*. Although such a summary can at most have only a provisional value, it is nevertheless of great assistance to the author, who is thus compelled to reduce his views to the conciseness of brief expression, to the reader who is thus enabled to see the results in a definite form, and to the future investigator, who thus finds a clearly circumscribed starting-point, and is in a better position to express the differences to which his own observations may lead him.

It is a matter of long experience that truth is only born in the conflict of opinions. If this maxim has proved itself to be correct in the descriptive sciences, how much more applicable will it be to a science which treats of causes!

Accordingly, the better to serve truth, the Archives will furnish space for the most conflicting opinions, provided they be supported by a *basis of observation.*

But *one* limitation is to be wished for in the approaching struggle, and it will be the endeavor of the editor to attain it in these Archives: the maintaining of a respectful tone even towards those who hold very different opinions. The ascertainment of truth, for which we are all seeking, is not furthered but retarded by the expression of personal feelings. Sufficient space will always be allotted to a proper treatment of differences and to remarks on priority.

The more vehement the struggle waged for the truth between different contentions, the more rapidly, generally speaking, shall we approach the lofty and distant goal of our ambition.

The specific processes of life are bound to the form and structure of its substrata. Hence *developmental mechanics* as the science of the causes of these formations will sometime constitute *the common basis of all other biological disciplines* and, *in continual symbiosis with these*, play a prominent part in the solutions of the problems of life.

At present opinions on the subject of developmental mechanics are much divided. While several biologists regard attempts in this direction as little more than the hobby of a few authors, and others are of the opinion that "so small a field" cannot pretend to maintain a publication of its own, the other conviction is already gaining ground that developmental mechanics is destined to become a science that will interest all the other biological disciplines.

That such will be the case is evinced in the most encouraging manner by the list of collaborators of these Archives — a list in which all the great departments of biology are represented. Besides these many other prominent investigators have expressed their interest and sympathy in the new tendency and in its organ. In this place I would again express my gratitude to all of these gentlemen.

INNSBRUCK, August, 1894.

LIST OF LITERATURE.

1. For more detailed information see: Roux, Wilh. Beitrag zur Entwickelungsmechanik des Embryo. *Zeitschr. für Biologie.* Bd XXI. Munich. 1885. (Separatum, p. 6.)

2. Roux, Wilh. Ziele und Wege der Entwickelungsmechanik, in Merkel and Bonnett's Ergebnisse der Anatomie und Entwickelungsgeschichte. 1892. Bd. II, p. 434.

3. Roux, Wilh. Ueber den Cytotropismus der Furchungszellen des braunen Frosches. (See the first article in the *Archiv für Entwickelungsmechanik.*)

4. See No. 2, p. 434.

5. Roux, Wilh. Ueber richtende und qualitative Wechselbeziehungen zwischen Zellleib und Zellkern. *Zoolog. Anzeig.* 1893. No. 432.

6. Roux, Wilh. Der Kampf der Theile im Organismus. Leipzig. 1881.

7. See No. 2, p. 435.

8. v. Sachs, Jul. Physiologische Notizen. No. 8. Mechanomorphosen und Phylogenie. *Flora od. Allg. Bot. Zeitung.* 1894. Heft 3.

9. See No. 1, p. 108.

10. Balfour, Francis M. Treatise on Comparative Embryology. (German translation by B. Vetter.) 1880. Vol. I, pp. 98–104.

11. See No. 1, p. 14.

12. Roux, Wilh. Die Methoden zur Erzeugung halber Froschembryonen und zum Nachweis der Beziehung der ersten Furchungsebenen des Froscheies zur Medianebene des Embryo. *Anat. Anzeig.* 1894. Bd. IX, Heft 8, p. 279.

13. Roux, Wilh. Ueber das entwickelungsmechanische Vermögen jeder der beiden ersten Furchungszellen des Eies. *Verhandl. d. Anatom. Gesellschaft zu Wien.* 1892. p. 57.

14. Conf. Dreyer, Friedr. Ziele und Wege biologischer Forschung, beleuchtet an der Hand einer Gerüstbildungsmechanik, p. 83. Jena. 1892.

15. v. Baer, Carl Ernst. Ueber Entwickelungsgeschichte der Thiere. Beobachtung und Reflexion. Theil I, p. 22. 1828.

16. See No. 1, p. 6.

17. Roux, Wilh. Ueber die Specifikation der Furchungszellen und über die bei der Postgeneration und Regeneration anzunehmenden Vorgänge. *Biol. Centralb.* Bd. XIII. 1893. p. 657 *et seq.*

18. His, Wilh. Ueber mechanische Grundvorgänge thierischer Formenbildung. *Archiv. f. Anat. u. Physiol. Anat. Abthg.* 1894.

19. Haeckel, Ernst. Generelle Morphologie der Organismen. Bd. II, p. 193 *et seq.* Berlin. 1866. *Idem*, Natürliche Schöpfungsgeschichte. 8. Aufl. Berlin. 1889. p. 212.

20. GEGENBAUR, CARL. *Morph. Jahrb.* 1876. Bd. I, p. 6.

21. AEBY, CHR. Der Bronchialbaum des Menschen und der Säugethiere. Leipzig. 1880.

22. WIEDERSHEIM, ROB. Grundriss der vergleichenden Anatomie der Wirbelthiere. 3. Aufl. Jena. 1893. p. 153.

23. FÜRBRINGER, MAX. Untersuchungen zur Morphologie und Systematik der Vögel. II. Allgemeiner Theil. Amsterdam. 1888.

24. LOEB, JACQUES. Untersuchungen zur physiologischen Morphologie der Thiere. I. Heteromorphosis. Würzburg. 1881.

25. NOTHNAGEL, H. Die Anpassung des Organismus bei pathologischen Veränderungen. *Wiener medic. Blätter.* 1894. No. 14. Vortrag gehalten auf dem internationalen medicinischen Kongress zu Rom.

4

EDWIN GRANT CONKLIN
1863–1952

Conklin received a B.A. from Ohio Wesleyan, then taught at Rust University in Mississippi. Thereafter, he attended graduate school in biology at Johns Hopkins, a close contemporary of Thomas Hunt Morgan and Ross Harrison, and received his Ph.D. in 1891. He taught at Ohio Wesleyan, Northwestern, and the University of Pennsylvania through the 1890s, then moved finally to Princeton in 1908. While in graduate school, Conklin went to Woods Hole to work at the Hopkins table at the United States Fish Commission. There he began to trace the early cell lineages in the gastropod *Crepidula*, despite criticism of his "cell counting" from his dissertation advisor, William Keith Brooks. At that time, Edmund Beecher Wilson was pursuing similar study of the annelid *Nereis* at the MBL and suggested that the two compare their results. They found quite remarkable homologies between the two forms. Both men shared a background of introduction at Johns Hopkins to the morphological problems and techniques of the day. Therefore, they both addressed questions of ancestral relations of their organisms and explored what the homologies they had discovered really told them about present or past relationships. The complex relation of present development to adaptation to either present or past conditions formed a central issue for their study. Both examined the nature of cleavage, of homologies, of heredity and evolution, and what each of these revealed about the others.

Conklin returned to Woods Hole as instructor at the MBL in 1892 and regularly attended the summer sessions thereafter, becoming a Trustee in 1896. This paper of 1896 explores the relation of cleavage to differentiation during development: Does cell cleavage cut apart a homogeneous mass of material, divide a heterogeneous mass randomly, or divide it according to some predelineated pattern. Conklin showed that there is no one simple answer, and then went on to distinguish different types of cleavage, thus clarifying what had been a very muddy pool of inconsistencies and contradictory alternative theories.

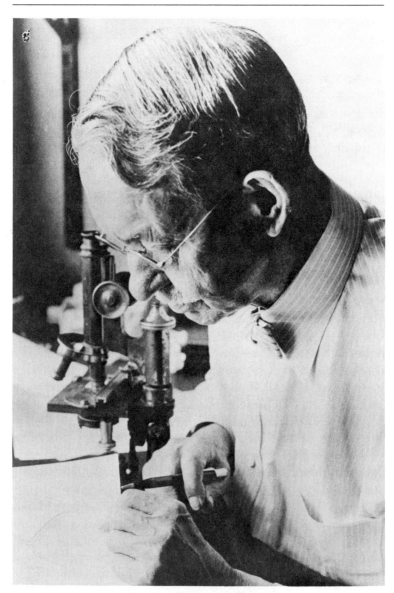

EDWIN GRANT CONKLIN.

CLEAVAGE AND DIFFERENTIATION

E. G. CONKLIN

PHILOSOPHICALLY, the most important problems of biology are those which concern the origin of a new individual, the genesis of a living organism. To the great problem of development has been devoted the earnest thought of philosophers and scientists of every age. The mystery which hangs about the process of progressive and coördinated differentiation by which the egg cell is transformed into the adult never loses its charm nor ceases to be a mystery.

Recent years have witnessed the most remarkable activity in this field, and the views now extant are so numerous, so difficult of concise representation, and have been so frequently discussed that it seems undesirable to dwell upon many of them here. In this lecture I shall present some observations and conclusions derived from a study of the normal development of certain gasteropods and shall attempt to apply these results to some of the current theories of development. Unfortunately, the nature of this material is such as to render direct experiment difficult and in most cases unsatisfactory. Observation, however, is still a valuable method in biology, and it has by no means revealed all that it can, either as to the course or the causes of development. It seems to be assumed in certain quarters that we already know all the important phenomena of normal development and that mere observation is, therefore, a useless and antiquated method. If the time ever comes when every step in the normal development of a single individual is known, the causes of development will not be far to seek. There is no such sharp distinction between observation and experiment in biology as is sometimes assumed; neither method can arrogate to itself a monopoly of certitude regarding facts or causes. In the solution of the problems of development both observation and experiment are necessary; each has its advantages and its disadvantages and one is no less important than the other.

I. DETERMINATE CLEAVAGE.

Without attempting any final and elaborate definition of so general a term as development, we may for our present purposes say that it is progressive and coördinated differentiation. In all Metazoa and Metaphyta the stages immediately following fertilization are characterized by the cleavage of the egg into a considerable number of cells. The question at once arises as to the relation between cleavage and differentiation. Is differentiation manifested in the cleavage of the egg? Is there any causal relation between cell-formation and differentiation?

There is abundant evidence that there is no *necessary* relation between the two. Many instances of differentiation without cell-formation might be given, *e.g.*, many Protozoa, Protophyta, the spermatozoa and ova of certain animals, intracellular differentiation of many tissue cells, etc. On the other hand, cell-formation may occur without differentiation, *e.g.*, all ordinary divisions of tissue cells and many divisions of embryonic cells. When the two processes are related we may have: (1) cell-formation following the lines of preceding differentiation, *e.g.*, certain cleavages of ctenophores, mollusks, and ascidians; or (2) cell-formation and concomitant differentiation, *e.g.*, many cleavages of turbellarians, nematodes, annelids, and mollusks; or (3) differentiation following the lines of preceding cell-formation, *e.g.*, many cleavages in the eggs of annelids, mollusks and probably many other animals.

In that pioneer work on developmental mechanics (*Unsere Körperform*, 1874) Wilhelm His propounded the doctrine that the organs and parts of an embryo are represented in the early stages of development, perhaps even in the unsegmented egg, by definitely localized germs (*Anlagen*). "The principle, according to which the germinal disk contains the preformed germs of organs spread out over a flat surface, and conversely, that every point of the germinal disk is found again in a later organ, I call the Principle of Organ-forming Germ-regions (*organbildende Keimbezirke*)." This doctrine has been denied in its totality by some authors, but, although it is still the subject of much controversy, the evidence is accumulating that with certain modifications it is true of a considerable number of animals belonging to several different types. The fact that, under unusual or "abnormal" conditions, regions which would have developed into certain parts develop into others is not a contradiction of the entire principle, though it does limit its causal significance.

Accepting the principle of His as true in certain cases, the relation of cleavage to these " germ regions " might conceivably be of two kinds; cleavage planes might follow the lines of separation between these regions, in which case there would follow a definite form of cleavage, each blastomere being destined to give rise to definite organs or parts of the embryo; or cleavage planes might cut across these regions indiscriminately, in which case an indefinite and inconstant form of cleavage would probably result. Of course, if one does not accept the principle of His, a third alternative is possible and is, in fact, imperative, *viz.*, cleavage is a mere sundering of homogeneous materials and every blastomere at the time of its formation is like every other blastomere.

The first of these alternatives has been presented in what is commonly called the " mosaic theory " of Roux;[1] the second in what might be called the " organization theory " of Whitman;[2] the third in what I venture to call the "homogeneity theory " of Driesch.[3] Disregarding for the present the *causes* of differentiation and viewing merely its *results*, it is probable that each one of these theories is true in certain cases. The study of cell-lineage has shown that in any given species among annelids, mollusks, ascidians, nematodes, and probably among ctenophores, turbellarians, rotifers, and crustacea each blasto-

[1] Roux, W., " Beiträge zur Entwicklungsmechanik des Embryo," Nr. V, 1888.

[2] Whitman, C. O., " The Inadequacy of the Cell-Theory of Development," Biological Lectures, Wood's Holl, 1893.

[3] Driesch, H., " Entwicklungsmechanische Studien," I–VI, *Zeit. wiss. Zool.*, Bde. 53, 55, 1891–93.

mere arises at a definite time, in a definite way, divides into a definite number of cells, each having definite characters, and in the end gives rise to a definite part. In such cases, as Wilson[1] has well said: " The development is a visible mosaic work, not one ideally conceived by a mental projection of the adult characteristics back upon the cleavage stages." Especially in the case of the annelids and mollusks the cleavage is a mosaic work more perfect than anything described by Roux, almost every organ of the larva being represented by a differentiated cell or group of cells before gastrulation is completed.

On the other hand, no such definiteness is known to exist in most cnidaria, echinoderms, and vertebrates, and is, in fact, denied by several excellent observers. In such cases the cleavage is equally inconstant, indefinite, and devoid of morphological significance, whether one conceives with Whitman that the unsegmented egg is mapped out into " germ regions," which are traversed in various directions by the cleavage planes, or whether one holds with Driesch that no such "preorganization" of the egg exists, and that " by cleavage perfectly homogeneous parts are formed capable of any fate."

Obviously the same considerations apply to the axial relations of the cleavage planes and, in case one denies the principle of His, to the polarity of the unsegmented egg. In all cases in which the cleavage has a mosaic character the relation of the egg-axis and of the planes of cleavage to the embryo or adult are perfectly definite and constant, and in many cases in which the cell lineage has not been followed and in which the mosaic character of the cleavage has not been directly recognized the constant relation of the planes of the first and second cleavages to the future planes of symmetry would indicate that the blastomeres bear constant relations to future organs. Whereas in those cases in which the egg-axis or the position of the early cleavage planes is inconstant the individual blastomeres can bear no constant relation to adult structures.

Confusion has already arisen through a failure to distinguish these two types of cleavage; much of the recent experimental

[1] Wilson, E. B., " The Mosaic Theory of Development," Biological Lectures, Wood's Holl, 1893.

work in embryology has been done upon forms in which the cleavage is not known to be constant, and general conclusions have been drawn which are plainly inapplicable to forms in which the cleavage is constant and definite. Although it is probable that there are forms which are intermediate between those which show extreme constancy and those which manifest extreme inconstancy of cleavage, yet the existence of two such *types* of cleavage must be recognized, and, as it is desirable to clearly distinguish between them, I propose to designate these types by the terms *determinate* and *indeterminate*. This is to be understood as applying only to the cleavage, for in its main features and results the development of all animals is determinate, that is, predictable. Even in cnidaria, echinoderms, and vertebrates the *general form* of the cleavage is constant and there appears successively a blastula, gastrula, larva, and adult of determinate form and character. The question is whether such determinism, which appears sooner or later in all cases, applies to the individual blastomeres of the cleavage stages.

Determinate cleavage is both *constant and differential*. It is more than constant, for in constant cleavage every blastomere might be like every other (Driesch); it is more than differential, for differential cleavage might be of such a sort that it is never twice alike (Whitman). It is the same as *mosaic cleavage*, but this name is not used because of the implication which it involves as to the cause of differentiation; determinate cleavage does not necessarily imply " self-differentiation " of blastomeres, which is such an important part of Roux's "mosaic theory." Cleavage is indeterminate when it is either inconstant or non-differential or both.

Among certain gasteropods [1] which I have studied the cleavage is of a highly determinate character as regards both the history and destiny of individual blastomeres and the relation of the cleavage planes and egg-axis to the future planes of symmetry. The chief axis of the ovum is established before fertilization, probably in the ovary, and it determines the

[1] Four species of Crepidula, Urosalpinx, Sycotypus, Fulgur, Tritia, Illyonassa, and Bulla.

gastrular axis and the cephalic and oral poles of the larva. In many cases the antero-posterior axis is marked out by the inequality of the first cleavage, and this is preceded by the eccentricity of the nuclear spindle, which in turn must be the result of the structure of the unsegmented egg. The direction of the first cleavage in Crepidula and probably in the other cases mentioned is always dexiotropic, that is, of such a character that the nuclei and protoplasmic areas of the two resulting cells rotate in a clockwise direction at the close of the cleavage (Fig. 1). This character must also be predetermined in the unsegmented egg. It is the first of a long series

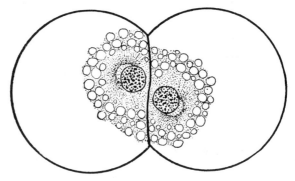

Fig. 1. — Crepidula, 2-cell stage; showing dexiotropic rotation of nuclei asters and cytoplasm at the close of the first cleavage.

of " spiral cleavages " (Figs. 2, 3, 4) which are oblique alternately to the right and to the left, each of which, except the first, finds the cause of its direction in that of the preceding cleavage. The direction of these cleavages stands in the most intimate relation to the origin of the mesoblastic pole cells, the appearance of bilateral symmetry, and the direction of the asymmetry of the adult. In all cases in which the first cleavage is dexiotropic the pole cells of the mesoblast arise from the left posterior macromere by laeotropic division (Fig. 4); where the first cleavage is laeotropic (as in some sinistral gasteropods) they arise from the right posterior macromere by dexiotropic division. In Crepidula bilateral symmetry appears in different directions in the ectoblast, mesoblast, and entoblast, and by a subsequent laeotropic rotation, which is dependent upon the direction of certain cleavages and ultimately upon the first cleavage, these diverse planes of symmetry come to coincide in a common plane. The direction of the asymmetry of the adult Crepidula is also referable to the time and direction of certain cleavages (of the fifth quartette) which are explained in part by the direction of preceding divisions and finally by the direction of the first cleavage; whereas in certain sinistral gasteropods, as Crampton[1] and Kofoid[2] have shown, the direction of all the cleavages is reversed.

All of these important and determinate characters are directly

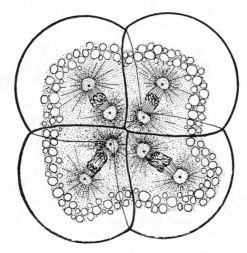

Fig. 2.—Crepidula, third cleavage; early indications of a dexiotropic rotation.

referable to certain peculiarities of the unsegmented egg, and although it is not possible to trace all determinate characters to this early stage, yet it is highly probable that many others are due to the same cause. How suggestive in this connection are the observations of Blochmann[3] upon the *Urvelarzellen* of Neritina; these cells contain a mass of coarse granules which can be traced back through previous generations of cells until

[1] Crampton, H. E., "Reversal of Cleavage in a Sinistral Gasteropod," Ann. New York Acad. Sciences, VIII, 1894.

[2] Kofoid, C. A., "On Some Laws of Cleavage in Limax," Proc. Am. Acad. Arts and Sciences, XXIX, 1894.

[3] Blochmann, F., "Ueber die Entwicklung der Neritina fluviatilis," *Zeit. wiss. Zool.*, Bd. 36, 1881.

they appear in the protoplasm of the unsegmented egg itself on each side of the animal pole. Likewise the observations of Driesch and Morgan[1] on ctenophore eggs indicate what a high degree of organization the unsegmented egg may reach. And while it is conceivable that this high degree of organization of the egg may not lead to a highly determinate form of cleavage, yet it is to be observed that in all the cases named this does happen.

All the earlier cleavages in Crepidula are spiral, that is, radially symmetrical, and this radial symmetry extends not only to the

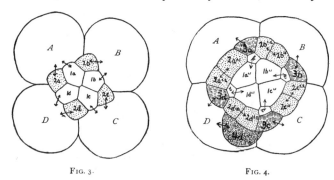

Fig. 3.   Fig. 4.

Fig. 3. — Crepidula, 12-cell stage ; four macromeres and eight micromeres.

Fig. 4. — Crepidula, twenty-five cells ; t, trochoblasts. In these and some of the following figures the macromeres and first quartette are unshaded ; the second quartette is stippled ; the third quartette is shaded with lines ; and the fourth quartette (4d) with dots and circles. The direction of the various cleavages is shown by means of arrows.

direction and time or rate of division, but also to the size, the position, and the histological character of the resulting blastomeres. The result is a number of radial structures such as the four trochoblasts (Fig. 4, t), the four arms of the ectoblastic cross (Fig. 5 et seq.), and the four rosette series of cells (Figs. 10, 12), some of which give rise to certain radial structures of the larva. Not a single bilateral cleavage appears up to the 44-cell stage, and radial cleavages generally prevail throughout the egg until a much later period. In all cases bilateral cleavages first appear in certain cells on the posterior side of the egg and in processes which lead to the elongation of the body along the posterior axis. This bilater-

[1] Driesch und Morgan, " Zur Analysis der ersten Entwicklungsstadien des Ctenophoreneies," Arch. für Entwicklungsmechanik, Bd. 2, 1895.

ality of the cleavage is directly and causally related to the bilaterality of the larva and the adult, though in some cases extensive rotations of cells and even of entire layers are necessary in order to bring blastomeres and planes of symmetry into their proper positions.

Apart from qualitative cell divisions, which are undoubtedly an important factor in differentiation, differential cleavages are the result of differences in the time and direction of division and in the size of the daughter cells. If divisions were always synchronous, alternating, and equal almost all the visible features of differential cleavage would disappear; it is in the constancy of certain peculiarities in the rate and direction of division and in the size of resulting cells that determinate cleavage is chiefly manifest.

Among the gasteropods mentioned above, the rate of growth and division of certain cells is highly peculiar, and in general this cannot be explained by the presence of yolk or by other extrinsic (that is, non-protoplasmic) causes. Adjacent and apparently homogeneous cells may behave in the most remarkably unlike ways in this regard. For example, the trochoblasts are at the time of their formation the smallest cells in the entire egg (Fig. 4); they grow rapidly, but divide rarely, and are characterized by having clear, non-granular protoplasm. On the other hand, the apical cells which gave rise to the trochoblasts are composed of granular protoplasm, and, although they grow scarcely more than the trochoblasts, they divide repeatedly, each of them giving rise at the stage shown in Fig. 10 to twelve cells, the total volume of which scarcely exceeds that of a single trochoblast. Many other illustrations of this same fact might be given.

In the departure of certain cells from the rule of alternating cleavage, or Sachs' law of rectangular intersection, we have another factor of differentiation and a marked feature of determinate cleavage. This is beautifully shown among the gasteropods named in the transition from radial to bilateral cleavages; in such cases the direction of division is reversed usually in one cell of a quartette (Fig 6). It is also shown in all cases of teloblastic growth, of which there are many at the posterior pole of the egg, where repeated divisions are in the same direction, and apparently in the shortest diameter of the protoplasm and in the line of greatest resistance. It appears also in the formation of certain definite structures, such as the ectoblastic cross, where the direction of a certain division is reversed in each arm. Upon this reversal depends the existence of the cross as such, and presumably of certain structures to which it gives rise.

Another remarkable instance of determinate cleavage is found in the unequal division of cells. Such unequal division constantly occurs in the formation of certain cells and is one of the most striking features of determinate cleavage. As has been said, the first cleavage may be unequal, though in most species the first and second cleavages divide the egg into nearly equal cells. In the formation of the three quartettes of ectomeres, however, the divisions are usually very unequal (Figs. 2, 3, 4), while in the formation of the fourth and fifth quartettes divisions are again more nearly equal. I have already called attention to the very small size of the trochoblasts when first

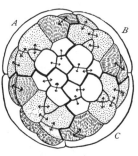
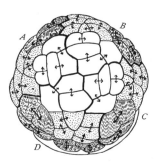

Fig. 5.   Fig. 6.

Fig. 5. — Crepidula, 42-cell stage. Shading as in Figs. 3, 4. The cross (shown in strong outline) lies in the position in which it was first formed. The heavy, radiating lines separate the cells of the different quadrants.

Fig. 6. — Crepidula, 60-cell stage. The whole of the ectoblast has rotated to the left, due to the rotation of the fourth-quartette cells. The middle cells in three arms of the cross have divided transversely. The third-quartette cells of the posterior quadrants have divided bilaterally.

formed; another illustration is found in the tip cells of the cross (Fig. 5). In fact, no phenomenon is more common in determinate cleavage than the unequal division of apparently homogeneous cells; such divisions are extremely constant and in many cases are visibly differential.   Even in the case of the echinoderm egg it has been shown that four micromeres are constantly formed at one pole of the egg, and in this respect, at least, the cleavage here is determinate, for although Driesch has shown that a normal larva develops from a sea-urchin egg from which the micromeres have been removed, this no more indicates, as Morgan [1] assumes, that these micromeres are undifferentiated and that the cleavage is, therefore, indeterminate than the fact that a hydra is able to complete itself and form a normal hydra after its tentacles

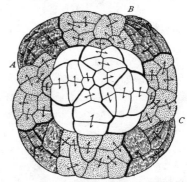

Fig. 7.— Crepidula, 109-cell stage (ninety-two ectoblast cells). Shading and heavy lines as in preceding figures. The egg is represented as if all the ectoblast cells could be seen from the apical pole, though actually many of the peripheral cells lie far down on the sides, or even on the ventral face of the egg.

have been removed indicates that these tentacles are undifferentiated.

The one most striking feature of determinate cleavage is the constancy with which certain blastomeres give rise to certain organs, the invariable segregation of an entire region, layer, or organ into a single cell or particular group of cells.   In all the gasteropods mentioned above the ectoderm comes from three quartettes of cells, each of which occupies relatively the same position and gives rise to the same organs (Fig. 4).   The mesoderm comes from the posterior cell of the fourth quartette.   All the other cells are entodermal, and, although they show certain variations in size and position in different genera, owing to variations in the amount and distribution of yolk, they are always constant for the same species.   The four apical cells give rise to an apical sense organ (see Figs. 3–10), the trochoblasts and tip cells of the cross form the first velar row, the anterior arm of the cross forms the anterior cell plate,

[1] Morgan, T. H., "A Study of a Variation in Cleavages," *Arch. für Entwicklungsmechanik*, Bd. 2. Hft. 1.

the posterior arm the posterior cell plate, the anterior rosette series gives rise (at least in part) to the cerebral ganglia, the shell gland and growing point come from the posterior member of the second quartette (2d), the paired mesoblast bands and the distal end of the intestine from the posterior member of the fourth quartette (4d), the roof of the archenteron from the remains of the four primary macromeres, its sides and floor from the fifth and fourth quartettes respectively; in fact, so many cells may be traced through to definite organs or parts that one is justified in concluding that under normal conditions every one of the earlier blastomeres gives rise to a particular part. *The constancy with which differentiated cells give rise to differentiated layers, regions, and organs is the most fundamental fact of determinate cleavage.*

What is the cause of determinate cleavage?

Such widespread, precise, and constant phenomena cannot, of course, be due to chance; nor are they the result of universally acting mechanical causes, such as gravity or surface tension. Certain indeterminate features of cleavage may be directly referred to extrinsic factors or mechanical conditions; *e.g.*, the rotation of cells into the furrows between blastomeres is probably referable to the principle of surface tension or mutual pressure, the contour of cells is frequently the result of intercellular pressure, the alternation of successive cleavages is an expression of the principle of rectangular intersection of cleavage planes, and this in turn may be due to the fact that the nuclear spindle usually lies in the direction of the greatest mass of protoplasm, and hence in the direction of least resistance. These features, however, are neither constant nor differential. So far as the principle of surface tension is concerned cells might rotate to the right or to the left indiscriminately, yet in determinate cleavage the direction of rotation is perfectly constant. So, also, it frequently happens that successive cleavages do not alternate in direction, and in such cases the nuclear spindles often appear to lie in the direction of greatest pressure. In general, the direction of teloblastic and non-alternating cleavages can be referred only to peculiarities in the protoplasmic structure of the cells, and, as I have pointed out, the constancy with which the first cleavage is dexiotropic is evidence of a constant peculiarity of the protoplasm of the unsegmented egg. Likewise the factors which determine the varying rate of division of certain blastomeres are generally intrinsic and protoplasmic rather than extrinsic; on no other basis can one explain the great difference in the rate of divi-

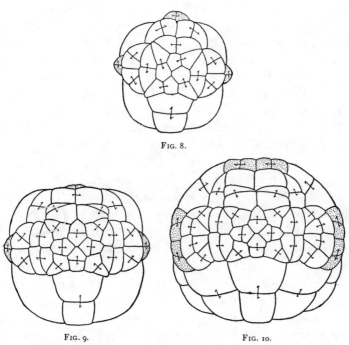

FIG. 8.

FIG. 9.   FIG. 10.

FIGS. 8-10. — First quartette in Crepidula, showing the later history of the cross and trochoblasts.

sion of contiguous cells. It is the same with that other marked character of determinate cleavage, — unequal divisions. In all cases in which unequal cleavage is not forced upon a cell from without, e.g., by unequal pressure, it must be regarded as an expression of a difference in the material substance of the dividing cell. In the separation of the micromeres from the macromeres there is a most marked material differentiation, one cell being purely protoplasmic, the other containing all the yolk. Even in cases of unequal cleavage in which the cell substance is apparently homogeneous, as, for example, in the formation of the trochoblasts and of the basal and tip cells in the arms of the cross, the initial eccentricity of the nuclear spindle indicates that here also there must be some difference of material substance within the cell, though not directly visible. Sachs[1] has well said, "The external form as well as the internal structure of any body are the necessary expression of its material constitution. Difference in form always indicates difference of material substance." That the cause of unequal cleavage is more complex than the mere mechanical displacement of the nuclear spindle is proven by the fact that the first two divisions of the egg are frequently equal, though the polar differentiation of the protoplasm and yolk is as marked as in later divisions which are very unequal.

What and how many factors enter into the complex of causes which produce even such simple phenomena as non-alternating, non-rhythmical, and unequal cleavages it is at present impossible to say; however, the prospective significance, the "purposefulness," of such cleavages is often very apparent. Lillie[2] has pointed out the fact that unequal cleavages in Unio stand in direct relation to the size of the parts arising from the blastomeres. With the following slight modification this principle is applicable to all the gasteropods which I have studied, viz., the initial size of the blastomere stands in direct relation to the size and the time of formation of the part to which it gives rise. In fact, this is but a partial expression of a much more general truth, viz., that all differential cleavages, whether non-alternating, non-rhythmical, or unequal, are directly and causally related to the uses to which these cells are put, — in short, to the general differentiation of the organism.

Other attempts have been made to explain the definite relation between blastomeres and organs than the one here given, viz., that the differentiation of the blastomere stands in direct relation to the differentiation of the parts and that the former is the result of differences in the material constitution of the cells. Hertwig[3] ascribes the fact that organs may be traced

---

[1] Sachs, J. v., "Physiology of Plants," Lecture XLIII, 1882.

[2] Lillie, F. R., "The Embryology of the Unionidae," *Journal of Morphology*, X, 1895.

[3] Hertwig, O., "Urmund und Spina-bifida," *Arch. für mik. Anat.*, Bd. 39, 1892.

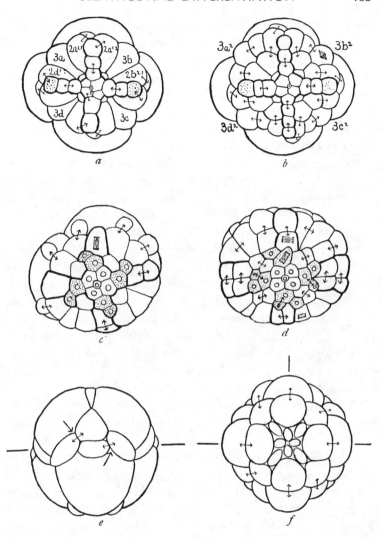

FIG. 11.—The cross in Neritina, Umbrella, and Chiton. — *a*, Neritina : three cells in each arm except the posterior; the granular tip cells of the transverse arms are the " Urvelarzellen." (Blochmann's Fig. 53.) — *b*, Neritina : four cells in the posterior arm, three in each of the others. The probable origin of the outer belt cells is indicated by arrows, and the designation of the cells in this and in the preceding figure are given as in Crepidula. (Blochmann's Fig. 56.) — *c*, Umbrella : the arms of the cross are stippled ; Heymons' so-called " cross " is shown in heavy outline. (Heymons' Fig. 14.) — *d*, Umbrella : stippling and outlines as in *c*. The basal cells in the arms of the cross have divided laeotropically, the trochoblasts bilaterally. (Heymons' Fig. 20.) — *e*, Chiton : lateral view of the 32-cell stage. The small cells around the equator of the egg correspond in origin and position to the *trochoblasts* and the *tip* cells of the gasteropod ; they should form the prototroch if they have the same destiny in the two cases. (Metcalf's Fig. XIV.) — *f*, Chiton : apical view of the 48-cell stage, showing the *cross*, the *rosette*, and the *trochoblasts*. (Metcalf's Fig. XXIV.)

back to certain blastomeres to the "continuity of development." " In consequence of the continuity of development," he says, " every older cell group must arise from a preceding younger cell group and so finally definite parts of the body from definite segment cells." A truer conclusion would be: and so finally definite parts of the body from any cell you please. Continuity of development no more explains the fact that the first cleavage is dexiotropic, that the ectoderm is segregated in three quartettes of cells, that the mesoderm comes from a definite cell of the fourth quartette, that certain cells always give rise to certain organs, than gravitation does. Likewise the "interaction of cells" which Hertwig and Driesch have invoked to explain so many features of differentiation is in this case an insufficient explanation. How can cellular interaction explain the fact that from the time of its formation a certain blastomere, *e.g.*, the *Urvelarzelle* of Neritina, is peculiar in size and histological character or that it grows rapidly and divides rarely, whereas an adjoining cell, the apical, grows slowly and divides rapidly ? If it is meant that differentiation is the result of the interaction of different material substances of the *protoplasm* which are more or less definitely localized, then there can, of course, be no objection to this view.

These are but a few of the many striking features of determinate cleavage which are not at present explicable by known mechanical causes. In the main one is compelled to refer determinism in development, whether it be in cleavage, the formation of organs, or the reproduction of specific and individual characters, to intrinsic causes, that is, to the structure of the germinal protoplasm, without for the present being able to explain *how* such protoplasmic structure is able to produce such predictable results.

Even Driesch, who represented very different ideas in his earlier writings, has said in one of his later papers:[1] "The facts make it necessary to suppose that there exists in the plasma-structure of every fertilized egg of a bilateral animal a polar-bilateral direction of its particles. . . . In addition there

[1] Driesch, H., " Betrachtungen über die Organisation des Eies und ihre Genese," *Arch. für Entwicklungsmechanik*, Bd. 4, 1896.

are present in many eggs different non-miscible substances which may predispose cells during cleavage to essentially different prospective values (micromeres and macromeres), and finally definite substances are definitely localized in the eggs of many

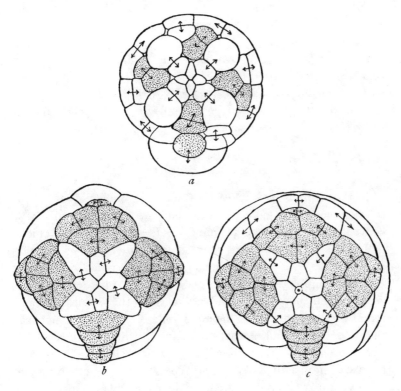

Fig. 12. — The cross in Nereis and Crepidula. — *a*, Nereis : the stippled cells are the *intermediate girdle cells* (molluscan cross) excepting the posterior one (x^3) which corresponds to the " tip cell " in the gasteropod. The trochoblasts lie at the margin of the egg. (Wilson's Diagram II. B.) — *b*, Crepidula : cross cells (intermediate girdle cells of Nereis) are stippled. Apical and rosette cells unshaded as in *a*. Trochoblasts around margin. — *c*, Crepidula : shading as in *b*; rosette cells and anterior trochoblasts divided.

animals which permit one to recognize necessary relations to certain early, firmly established organs. . . . In certain cases axial relations may be stamped upon eggs through the action of external factors; in the majority of cases, however, especially in eggs with complicated structure, this is not the case; the organization is here performed in the unfertilized egg, that is, it has arisen in the course of ontogeny (oögenesis) as a typical differentiation, at a typical place of the entire germ, through typical formative internal stimuli " (p. 99). After such sweeping concessions from the most vigorous opponent of the principle of His and of the mosaic theory of Roux we may now consider determinate cleavage, at least in certain cases, as no longer a matter of controversy. In conclusion one may say of all determinate cleavage that *the reason that a certain blastomere arises in a certain way, passes through a definite developmental history, and in the end gives rise to a definite part is at bottom the same reason that the egg of a given animal passes through a definite history and gives rise to a definite organism.*

II. Cell Homology.

In the search for the earliest appearing homologies in the development of organisms embryologists have generally been content to stop with the germinal layers. This has been chiefly due to the fact that there is such great diversity in the pregastrular stages of most animals that they cannot be brought into any single system. There are various types of cleavage, such as the meroblastic and holoblastic, the alecithal, telolecithal, and centrolecithal, the radial, bilateral, and asymmetrical, the determinate and indeterminate, and, while it is possible to hypothetically connect them, it is not possible at present to compare the blastomeres of one type with those of any other. If any similarity ever existed between the blastomeres of an arthropod and of an annelid or of a cephalopod and of a gasteropod the alteration of the type of cleavage has completely destroyed it. Any attempt to establish cell homologies must be limited not only to a single type, but also to determinate, that is, constant and differential cleavage. In addition, any detailed comparison of the cleavage stages of various animals demands an accurate knowledge of the cell-origin of various parts and organs, and this is unfortunately lacking except in a few cases.

If, within the limits indicated, we compare the cleavage of one species with that of other related species or genera we find many identical characters running through all of them. Among the gasteropods these resemblances of the cleavage stages are marvelously accurate and complete; even among forms showing the greatest diversity in the size and structure of the egg, in the method of gastrulation and in the adult these resemblances are minute and long continued.   Among the most diverse types of prosobranchs, opisthobranchs, and pulmonates very many blastomeres are identical in method of origin, relative position, and ultimate destiny.   In fact, so far as now known, all gasteropods have not only the same type of cleavage, but all manifest the most fundamental similarity in the developmental history of individual blastomeres.

The amount and distribution of yolk has little influence on these resemblances.   The egg of Crepidula adunca is 27 times as large as that of C. plana, and yet every cleavage is identically the same up to the 52-cell stage.   The egg of Fulgur is 140 times as large as that of C. plana, and yet the early cleavages and the ultimate fate of the blastomeres is almost exactly the same in the two cases.

In the distribution of the yolk the most diverse conditions are found associated with the most fundamental resemblances in the origin and history of the blastomeres.   In many eggs the yolk is equally distributed to the first four cells, *e.g.*, four species of Crepidula, Neritina, Planorbis, Sycotypus, Fulgur, and Bulla.   In others it is chiefly aggregated in one, two, or three of the macromeres, *e.g.*, Urosalpinx, Illyonassa, Tritia, Aplysia, Umbrella, etc.   In general, if one macromere is larger than another, it is the posterior one among prosobranchs and the anterior one among opisthobranchs.   Although this unequal distribution of yolk makes marked changes in the form of the embryo, it scarcely influences in a single respect the typical formation and development of blastomeres.

In one respect there seems to be a notable difference between forms otherwise remarkably alike.   In a large number of gasteropods (Neritina, Planorbis, Vermetus, Aplysia, Urosalpinx, Tritia, Nassa, Illyonassa, etc.) the first and second cleavage planes are oblique to the median plane of the embryo, whereas in another series of forms (Crepidula, Umbrella, Sycotypus, Fulgur, etc.) the first cleavage is approximately transverse to the median plane and the second coincides with it. The axial relations of the first two cleavages being different in these cases, it seems that the first four cells must give rise to different organs in the two classes named. However, a careful examination shows that in all these cases the ectomeres and mesomeres rotate so as to occupy relatively the same positions and ultimately give rise to the same parts (Fig. 6); the position of the entomeres alone is different. It seems to me very probable, considering the extensive shifting which the entomeres undergo in late stages, that even the axial differences of these cells may ultimately disappear, but even if they do not it is certainly a matter of secondary importance that a few cells forming a tubular internal canal should occupy slightly different axial relations as compared with the fact that hundreds of cells occupy relatively the same positions and give rise to the same organs. The entomeres have undergone great modifications owing to the acquisition and loss of yolk and its varying distribution to the different macromeres, and it would not be surprising if they have also shifted their axial relations in some cases. On the whole, this apparent difference in the axial relations of the first two cleavages affords an unexpected confirmation of the fundamental likeness of all gasteropod cleavage.

These important resemblances of cleavage stages are not limited to the gasteropods. Wilson [1] has pointed out a number of remarkable similarities in the cleavage of polyclades, annelids, and gasteropods; Lillie [2] has shown that the lamellibranch cleavage is essentially like that of the gasteropods and annelids; and Heath [3] has recently discovered that the cleavage of Chiton resembles in the most wonderful manner the cleavage of all the groups just named.

" Wilson emphasizes the following important resemblances between the early cleavage stages of the annelid, the polyclade, and the gasteropod: (1) the *number and direction of the cleavages* is the same in all three up to the 28-cell stage; (2) in

---

[1] Wilson, E. B., " The Cell Lineage of Nereis," *Journal of Morphology*, VI, 1892.
[2] Lillie, F. R. " The Embryology of the Unionidae," *Journal of Morphology*, X, 1895.   [3] Heath's work is not yet published.

general, the cells formed are *similar in position and size, viz.*, there are four macromeres, three quartettes of micromeres, and the first quartette is surrounded by a belt composed of the second and third quartettes. The first quartette undergoes three spiral divisions in alternate directions, and the second quartette divides once. Here the resemblance with the polyclade ceases, though the annelid and gasteropod go one step further in these likenesses, *viz.*, (3) the *three quartettes of micromeres are ectomeres* in the annelid and gasteropod, and (4) in both these groups *the mesoblast is formed from the cell 4d*, which gives rise to paired mesoblastic bands.

" Beyond this point Wilson believed that the annelid diverged from the gasteropod. He supposed that the ' cross ' in the two was wholly different both in origin, position, and destiny, and that the velum had a wholly different origin from the annelidan prototroch.

" Lillie has extended all the above-mentioned resemblances between annelids and gasteropods to the lamellibranchs, and in addition has discovered the following : (5) the *first somatoblast* (2d), which gives rise to the ectoderm of the trunk, has exactly the same origin and position and a similar history in the annelid and lamellibranch; (6) it gives rise to a *growing point* and a *ventral plate* in all respects essentially like those of the annelids. Lillie shows good reason for believing that in other mollusks the posterior growing point is derived from these cells.

" To this list of resemblances between the annelid and the mollusk, which I can confirm in the case of the gasteropod, I have been able to add the following: (7) the *rosette series* of the gasteropod is exactly like the *cross* of the annelid in origin, position, and probably in destiny. The *intermediate girdle cells* of the annelid are like the *cross* of the gasteropod in origin, position, and destiny (at least in part) (Fig. 12). The differences, therefore, between the annelidan and molluscan cross which Wilson emphasizes are not real ones; (8) the *trochoblasts* of the annelids and gasteropods are precisely similar in origin and destiny (at least in part) (Figs. 10, 12). In some annelids (Amphitrite, Clymenella, Arenicola), the prototroch is completed by cells of the same origin as in Crepidula and Neritina. The differences which Wilson points out between these two structures do not, therefore, exist. In both annelids and mollusks the prototroch lies at the boundary between the first quartette on one side and the second and third on the other. In both there is found a preoral, an adoral, and a post-oral band of cilia; (9) in the gasteropod the apical cells give rise to an *apical sense organ* such as is found in many annelid trochophores; (10) the *supra-oesophageal ganglia and commissure* apparently arise from the same group of cells in annelids and gasteropods; (11) the *fourth quartette* in annelids and gasteropods contains mesoblast in quadrant D, but is purely entoblastic in quadrants A, B, and C; (12) a *fifth quartette* is formed in gasteropods and some annelids (Amphitrite, etc.), and consists of entoblast only; (13) in the gasteropod *larval mesoblast* arises from the same group of ectoblast cells as in Unio, differing, however, in this regard, that it is found in quadrants A, B, and C, whereas in Unio it is found in quadrant A only; (14) to this list of accurate resemblances in the cleavage cells may be added the fact that *among annelids and mollusks the axial relations of all the blastomeres (except possibly the four macromeres) are the same.*

"What a wonderful parallel is this between animals so unlike in their end stages! How can such resemblances be explained? Are they merely the result of such mechanical principles as surface tension, alternation of cleavage, etc., or do they have some common cause in the fundamental structure of the protoplasm itself? Driesch answers: 'The striking similarity between the types of cleavage of polyclades, gasteropods, and annelids does not appear startling; it is easy to understand this, since cleavage is of no systematic worth.' To this, I think, it need only be said in reply that if these minute and long-continued resemblances are of no systematic worth, and are merely the result of extrinsic causes, as is implied, then there are no resemblances between either embryos or adults that may not be so explained. And, conversely, these resemblances in cleavage, however they have been produced, stand upon the same basis with adult homologies."[1]

[1] "Embryology of Crepidula," *Journal of Morphology*, XIII, No. 1.

The cause of such resemblances, like the cause of determinate cleavage and of the constancy of specific characters, must be found in protoplasmic structure, and I cannot escape the conviction that these likenesses belong to the same category with the fundamental resemblances between gastrulae, larvae, and adults. Whatever criterion of homology one may adopt — whether similarity of origin, position, history, or destiny, or all of these combined — certain of these resemblances in cleavage bear all the marks of true homologies.

It is freely granted once for all that even in the limited form in which it is here maintained there are serious difficulties in the way of the doctrine of cell homology. The most important of these difficulties are the following: (1) Related animals do not always have similar cleavage, *e.g.*, cephalopods and other mollusks ; triclades, and polyclades. Even within a single order there may be important differences; thus the cleavage is markedly radial in Discocoelis and as markedly bilateral in Polychaerus. Among the Crustacea there are four types of cleavage (see Korschelt und Heider, *Lehrbuch der Entwicklungsgeschichte*): (*a*) total and equal, (*b*) total and later superficial, (*c*) purely superficial, (*d*) discoidal. Finally, contradictions reach a climax among the Daphnidae, where the summer and winter eggs of the same species may belong to wholly different types of cleavage, as Watasé[1] has pointed out. No cell homology is recognizable in such cases, and possibly none exists. (2) Similar larval or adult parts may arise through very different types of cleavage; *e.g.*, the primitive streak of sauropsida and mammalia, the adult structures of amphioxus as compared with most other vertebrates, the shell gland of gasteropods and cephalopods. Such cases show that adult homologies are not necessarily dependent upon cell homologies. (3) Similarities in cleavage may not lead to similarities in subsequent stages, *e.g.*, the cleavage of certain polyclades is closely like that of annelids and mollusks, and yet the cells which are mesomeres in one case are ectomeres in the other. However, the discovery of larval mesenchyme in Unio and Crepidula has lessened the difference in this regard, and it is possible that a further

[1] Watasé, S., " Studies on Cephalopods," *Journal of Morphology*, IV, 1891.

comparison would bring these two groups into still closer agreement. (4) Finally, experiment has shown that the form of determinate cleavage, which alone is under consideration, may be modified in certain regards without materially modifying the results of development. It must not be supposed, however, that such experiments destroy belief in either determinate cleavage or cell homology. That certain forms of cleavage are determinate, *i.e.*, under normal or usual conditions constant and differential, is a visible fact; that certain cells in related animals normally give rise to the same parts is also a fact which cannot be denied. Experiment shows that this normal condition may be modified; it does not prove its non-existence. Even if it should be shown that the apical organ might be formed in the absence of the apical cells or that the mesoblast might appear after the removal of the cell 4d — and be it observed such a thing has never been proved — the case would not be fundamentally different from the regeneration of adult parts after their complete loss, and the doctrine of homology would no more be destroyed in the one case than in the other. On the whole, experiments on determinate cleavage (*e.g.*, Driesch and Morgan[1] on the ctenophore and Crampton[2] on the gasteropod) lend support to the doctrine of cell homology.

A consideration of these difficulties, especially of the first and second, shows how futile is any attempt to establish the *universal* homology of blastomeres, and it indicates, as Wilson has pointed out in his lecture on the "Embryological Criterion of Homology," that embryological likeness or unlikeness is not in itself a sufficient test of homology; it indicates, as do many other considerations, that the early stages of development have undergone profound modifications in the course of evolution, but it does not prove that these early stages never resembled each other or that no traces of such primitive resemblance can now be found between related organisms. In all respects the same objections as those presented above may be urged against the homology of many embryonic structures and processes.

[1] *Op. cit*, p. 24.

[2] Crampton, H. E., "Experimental Studies on Gasteropod Development," *Arch. für Entwicklungsmechanik*, Bd. 3, 1896.

Numberless instances are known in which homologous adult parts arise in different ways in closely related animals—*e.g.*, the central nervous system of teleosts and of selachians, the notochord and mesoblastic somites of amphioxus and of other vertebrates, the body musculature of Lopadorynchus and of other annelids, etc.—and yet who holds on this account that there are no homologies whatsoever between any embryonic parts? The objections to such homologies are objections only to the view that they are complete and universal; among certain phyla and recognizing certain modifications, even the germ layers are homologous, and within perhaps even narrower limits there is homology of blastomeres. How else is it possible to explain the remarkable resemblances which have been pointed out between the annelids and mollusks, resemblances which are inherited with such tenacity as to be found throughout all the species, genera, and orders of an entire phylum? The fact that blastomeres are not universally homologous should not cause us to shut our eyes to certain striking homologies which do exist. Certainly within the limits here indicated the existence of cell homologies seems extremely probable, and their importance will not be overlooked save by those who are concerned only with "universal laws."

If such resemblances between blastomeres are homologies, what follows? (1) Cleavage has a certain phylogenetic significance, and, although possibly more liable to modifications than larval or adult stages and hence less trustworthy as a test of homology and of genetic relationship, it may in certain cases at least preserve ancestral conditions even after they have disappeared in end stages (annelids and mollusks). Incidentally, the homologies of cleavage added to those of embryonic and larval structures indicate the close relationship of annelids and mollusks, whereas the entire embryological history only serves to widen the gap between the cephalopods and other mollusks.

(2) The early cleavages are morphologically more important than later ones. This follows from the notion of determinate cleavage, some of the earlier blastomeres being destined to form entire regions or organs of the animal, but principally

from the fact that the earlier cleavages are more constant than the later ones. In all gasteropods, lamellibranchs, and annelids, so far as known, the early cleavages are almost identically the same; but in later stages there are certain differences in the cleavage of various species and genera, many additional cells, for example, being found in large eggs which are not found in small ones. Thus, whatever the size of the egg, three and only three quartettes of ectomeres are formed, which in all cases occupy relatively the same positions and give rise to the same organs. This is a fact of the widest application and of the highest significance; it occurs in equal and unequal cleavage and in eggs varying in size from a few microns to more than a millimeter in diameter. However, in the subdivisions of these quartettes marked differences sooner or later appear. In Crepidula plana, fornicata, convexa, and adunca the relative volumes of the eggs are as 1, $2\frac{2}{5}$, $8\frac{3}{4}$, $27\frac{2}{5}$, and yet up to the 52-cell stage there is not a single difference in the cleavage of these four species; but at this stage a single additional ectoderm cell appears in the large egg of C. adunca, due to the additional subdivision of one of the ectomeres; at the 82-cell stage there are three additional ectomeres; at a similar stage all the other species have the same number of cells, that is, three less than adunca, but in later stages the ectoderm cells divide more rapidly in all the large eggs than in the small ones, for at the time of the closure of the blastopore the number of ectoderm cells in the four species, plana, fornicata, convexa, and adunca, are in the following ratio: 1, 1.6, 2.6, 5. Finally, in the adult condition these proportions are reversed, the largest egg giving rise to the smallest individual with the smallest number of cells.

This difference in the number of cells offers no difficulty to the doctrine of cell homology unless we assume that all divisions are differential, a thing which we know is not true. After blocking out the protoblasts of various regions and organs an indefinite number of non-differential divisions may occur either before or after the complete differentiation of the parts, and this probably explains the larger number of cells in the embryo of C. adunca and the smaller number in the adult. In fact, after the complete differentiation of all the tissues and organs, the number of cells may vary greatly in different individuals of the same species or in the same individual at different times. In adult Crepidulas the number of cells varies directly as the body size varies, the cell size remaining practically constant. These later divisions, in the main, are non-differential, and likewise it is probable that in the later stages of cleavage many non-differential and inconstant divisions occur. Not only is there greater variation in the number and size of cells in later as compared with earlier stages of cleavage, but there is also greater variation in the direction and time of division; all of which goes to prove that the earlier cleavages are more constant, more frequently differential, and therefore morphologically more important. This view, though reached by a different line of reasoning, is in entire agreement with Watasé's [1] conclusions, and is opposed to those of Wilson.[2]

At first thought it may seem strange and improbable that the earlier cleavages should be more important than the later ones. It is generally, and I think truly, believed that processes of differentiation increase in extent as we approach the end stage. However, the greater differentiations of later stages are dependent upon the lesser differentiations of earlier ones, which are therefore causally the more important. Moreover, the later differentiations in general are not phenomena of individual cells, but of cell aggregates, whereas the differentiations of cleavage are primarily differentiations of individual cells. The mosaic character of cleavage is, therefore, most pronounced in early stages, whereas the cellular phenomena of differentiation become less prominent as development advances.

[1] *Op. cit.*, p. 38.
[2] *Op. cit.*, p. 36.

5

CORNELIA MARIA CLAPP
1849–1934

As a student at Mount Holyoke, Clapp studied with Lydia Shattuck, who attended Agassiz's Penikese school in 1873. The second year, Clapp went along with Shattuck, an experience which she regarded as opening "a thousand new doorways." She determined then to pursue biology and to teach by using the animals themselves rather than textbooks. She received a Ph.B. from Syracuse University, attended walking trips to collect in the White Mountains, through the southern states, and through Europe, spent five weeks studying with Edmund Beecher Wilson at Williams College, then attended the MBL's first session. An energetic woman with practical short hair and a willingness to wade in after an enticing specimen, Clapp was popular at the MBL from the beginning. She decided that she wanted a Ph.D. and received her degree under Whitman from the University of Chicago in 1896 for her work on toadfish development. She taught creatively and successfully at Mount Holyoke throughout her career and returned to the MBL virtually every summer until her death.

In 1901 Clapp became the first woman elected to the MBL Board of Trustees. She always supported Whitman in his advocacy of both instruction and investigation. Her direct style is revealed in her lecture to the MBL in 1898 on the question of whether the first cleavage determines the plane of the embryo. Pflüger, Roux, and Born had maintained that it did, and their experimental results had gained considerable attention even among those Americans skeptical of their interpretations. In keeping with the MBL community's interest in cleavage and differentiation, Clapp examined the significance of the first cleavage for the toadfish and discovered that Pflüger's and Roux's results did not hold. Neither could their interpretations. With typical spirit, Clapp insisted that their frogs' eggs proved inferior for their purpose to her toadfish eggs. Her results lent support to the view common at the MBL that cleavage was not generally as determinate as Roux demanded.

CORNELIA MARIA CLAPP.

RELATION OF THE AXIS OF THE EMBRYO TO THE FIRST CLEAVAGE PLANE

CORNELIA M. CLAPP

MT. HOLYOKE COLLEGE, SOUTH HADLEY, MASS.

WHAT is the meaning of cleavage? is the constantly recurring question of the embryologist. Is cleavage a differentiating process in development, or is it a process which may run on independently of differentiation, sometimes coinciding with it, sometimes not?

Is the egg a mosaic, as Roux maintains? Does the first plane of cleavage determine the axial position of the embryo, or is the egg already definitely oriented with respect to the future embryo before cleavage begins? Is the egg an isotropic body to be converted gradually into a definite mosaic as cleavage advances, or are the blastomeres all endowed with the same potentialities, the fate of each being settled by the position it happens to hold?

Does the division of the egg by the first line of cleavage mean the separation of the parts destined to become the right and left sides of the bilaterally symmetrical animal, or may this first cleavage run in any direction, and have no fixed and necessary relation to the future embryo?

The well-known discoveries of Newport ('54) and of Pflüger and Roux in '87 and '88 seem to point toward some general law controlling the phenomena of cleavage.

In 1891, while studying the development of *Batrachus tau* at the Marine Biological Laboratory at Wood's Holl, Professor Whitman directed my attention to the fact that the egg was a favorable one for testing the conclusions of Roux. An egg provided with an adhesive disk, by means of which it is held constantly in a fixed position, seemed to offer just the condition required for testing the mosaic theory of cleavage.

The results have been reported in the *Journal of Morphology*, vol. v, p. 498. In only *three* cases out of twenty-three was there coincidence of the first line of cleavage with the median plane of the embryo.

In 1892 Born[1] called attention to my paper and pointed out what he regarded as the "sources of error" in the case. His criticism was accepted by Roux,[2] who adds : "Only slight errors of experiment are required in order to obtain almost equally incorrect figures in the frog's egg."

Born doubts the validity of the results, because of the long period of time that elapses between the appearance of the first furrow and the outline of the embryo. During this period of six days, slight changes may occur to account for the "abweichende Resultat." Born also finds difficulty in accepting the assertion that "the adhesion of the yolk to the egg-membrane prevented rotation," and says that "such adhesion would certainly hold good only for the first stages."

Weysse,[3] in 1894, believing that if there was adhesion of the yolk to the membrane it could be demonstrated in sections of the egg, made a careful examination, but found no trace of any such attachment. He concludes, with Born, that "during the six days mentioned abundant opportunity is furnished for a rotation of the yolk within the egg-membrane."

These experiments have been twice repeated since '91, with essentially the same result. In order that the possible sources of error may be fairly estimated, it seems desirable, in giving the results of my later experiments, to repeat and perhaps expand the brief description of the toadfish egg given in '91.

The egg is large, — 5 mm. in diameter, — and as the blastodisk always develops at the free pole of the egg, the early cleavage lines can be observed with a hand-lens without dis-

[1] Merkel and Bonnet, Erg., Bd. I, p. 502.

[2] "Beiträg zur Entwickelungsmechanik des Embryo," *Anat.*, Heft 7, p. 313.

[3] Proc. of Amer. Acad., vol. xxx, p. 308.

turbing the egg. The egg is hardy, and develops well in shallow dishes holding sea-water about two inches deep.

The membrane of the egg has a peculiar adhesive disk, about 3 mm. in diameter, which has a constant position, with the centre of the disk at the vegetative pole, directly opposite the micropyle. By means of this disk the egg is firmly glued to the supporting surface, usually the underside of a rock, the inside of a broken jug, tin can, a piece of stovepipe, or even an old boot-leg. The disk consists of a transparent secretion, which becomes opaque and gluey on contact with water. It is of nearly uniform thickness, and is closely applied to the egg-membrane everywhere except for a narrow margin which pro-

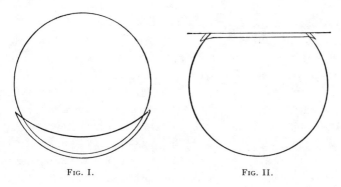

FIG. I.   FIG. II.

jects all around as a thin rim (Fig. I). The disk is saucer-shaped, and only a little thicker than the egg-membrane itself. I have been able to separate it from the membrane in the case of eggs hardened before attachment. As the egg is generally fastened to more or less plane surfaces, it appears strongly flattened on the side of attachment, as described by Dr. Ryder and as shown in Fig. II.

That the egg-membrane, with its adhesive disk, does not change position after attachment is, then, very certain; but is the egg itself also fixed within the membrane, and does it maintain a fixed position for the time required to reach an early stage in the axial differentiation — six or seven days after fertilization? or is it liable to rotate and get displaced, and so invalidate the results obtained?

In the experiment of '91 the eggs were allowed to flow from

the opened ovary of the fish into a glass dish containing just water enough to cover them. After becoming fixed to the bottom they were fertilized. The side of the egg resting on the glass was much flattened, and the fact that the blastodisk retained its position at the free pole of the egg when that pole was directed upward, led to the statement that "the adhesion of the yolk to the egg-membrane, as it rested on the disk area, prevented rotation."

In the experiments of '96 and '98 the eggs were allowed to fasten themselves on pieces of glass which were inverted and which rested on supports within a large dish of sea-water. Consequently the cleavage and development had to be observed from below, by looking up through the bottom of the glass dish.

In this case the eggs were in their natural position, that is, in the position in which they are usually deposited by the fish when making her nest under a flat stone.

It makes no difference, then, how the egg is attached, whether on the roof of the nest, or on the side or on the floor ; the axis of the egg is always *perpendicular to the plane of attachment*. The fish usually fixes its eggs so that they hang from the surface above ; but if they are deposited in a piece of stove-pipe, so as to cover the upper half of the concave inner surface, the position of the axis of the egg will vary at all angles between a vertical and horizontal position.

If the eggs are dropped loosely into a glass dish containing very little water, some of them will fall so as to become quickly fixed to the bottom of the dish, while others may fall so that the adhesive surface fails to come into contact with the glass, and so soon loses its adhesive property, leaving the egg free to roll. The axis of the fixed eggs will be vertical, but the free pole will face upward instead of downward, as it does when the egg is placed on the roof of the nest, or on the underside of a glass plate. It is evident, therefore, that gravity has no decisive influence in determining the direction of the egg axis. In other words, the egg has a definite and constant position in its membrane, the vegetative pole always lying at the centre of the adhesive disk, while the animal pole and micropyle coincide on the opposite side.

This is not saying that the egg cannot be forced to change its position within the membrane. The eggs that fail to become attached to the bottom of the dish are liable to be rolled about, and the movement may be sufficient to cause some shifting in position. The important fact here is, that, whatever be the place of attachment, whether above, below, or on the side, the axis of the egg always maintains the constant relation of being perpendicular to that point, and there is no inherent tendency to assume a *vertical* position, as is the case in most pelagic fish eggs.

There is, then, no cause for rotation within the membrane unless it be supplied artificially from without. The egg remains in the position in which it becomes fixed, neither rotating nor becoming displaced except as the result of rough treatment.

In the experiment of '98 great care was taken to guard against any possible disturbance of the eggs. The pieces of glass with the adherent eggs were arranged conveniently for observation before they were fertilized, and the water changed from time to time by means of a siphon, so that no disturbance seemed possible.

If the egg does not rotate during the early stages of development, it may be asked, What is the *nature* of the adhesion that prevents rotation?

In regard to this matter Professor Whitman assures me that adhesion of the yolk to the membrane is a very general phenomenon which he has noticed in the egg of *Necturus, Amblystoma,* frog, newt, pelagic fish eggs, and even in the small eggs of many annelids. Speaking of pelagic fish eggs, he says: " It is only necessary for an egg to be left at rest for a few minutes in order to see how readily it adheres to its membrane. I have often taken advantage of this adhesion to roll over and hold them some moments by the aid of needles, with the lower pole uppermost. Others, I am sure, have done the same thing while studying the cleavage." He also adds that in the case of the toadfish "the strength of the adhesion is probably due to the relatively large weight of the egg and the large surface of contact." The perivitelline space in this egg is so slight as to be difficult of recognition, and the extent of the surface in actual contact with the membrane is correspondingly very great as compared with small eggs with ample space between the yolk and the membrane. The conditions, then, are most favorable for strong adhesion.

Born asserts that "such an adhesion cannot hold for more than the first stages." As this statement does not rest upon observation of the toadfish egg, it can have no value beyond that of conjecture suggested by experience with other eggs. It is quite true that six days gives *ample time* for any number of rotations, but it is not a question whether there is *time* for rotation, but *whether rotation actually occurs*, and that can only be determined by close study of the developing egg, not by sectioning of hardened and imbedded eggs, nor by any amount of experiments on frogs' eggs.

In one very important respect the toadfish egg is far superior to the frog's egg for the study of the question here considered, since it can be observed under perfectly normal conditions, without resort to those artificial means of fixation or marking which are necessary in a frog's egg, and which must always cast some doubt on the reliability of the results.

In the experiment of '91 the method of determining the relation of the first cleavage plane to the axis of the embryo was as follows: After the eggs had become fixed by the adhesive disk to the bottom of a glass dish they were artificially fertilized. The blastodisk appeared on the free pole of the egg, where it was easily watched by means of a lens. The eggs were plotted on paper, each egg being represented by a circle (Fig. III), and the paper and the dish containing the eggs oriented by fastening a label on each in the same relative position. When the first line of cleavage appeared the direction was indicated in the circle representing the egg, by the diameter, and when the axis of the embryo became visible, that was indicated by an arrow drawn across the same circle.

The first furrow appeared seven hours after fertilization, and on the seventh day the axis of the embryo could be distinctly seen as a light streak in the blastoderm.

The result of this experiment is seen by an examination of the circles given in Fig. III. Of the twenty-three developing embryos, *three* show coincidence of the axis of the embryo with the first cleavage plane.   There is no case of exact coincidence with the second cleavage plane.   Fourteen of the embryos have the head directed towards the right of the first line of cleavage, the axis of the body being at an angle with the first cleavage plane of from 30° to 70°.   In the remaining six the head was to the left of the first cleavage plane, the angle varying as before.

In the experiments of '96 the main purpose in view was to settle the question of rotation of the egg within its membrane

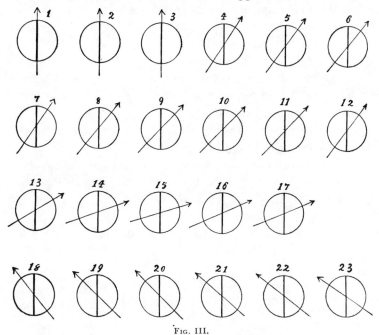

Fig. III.

and to find out, if possible, whether gravity had anything to do with the position of the blastodisk.   The eggs attached to pieces of glass were fertilized and the glass plates placed in various positions, so that the axis of the egg in different cases was vertical, horizontal, and inclined at various angles.

It was observed that the small oil globules within the egg were always found uppermost in whatever position the egg was placed, in one case being directly under the blastodisk at the animal pole, and again directly underneath the adhesive disk at the vegetative pole, or halfway between these poles *on the side that was uppermost.*

As for the blastodisk itself, *that retained its position at the free pole of the egg, near the micropyle, in whatever position the eggs were placed.*

The position of the eggs was plotted in the case of ten eggs attached in the inverted position.

The coincidence of the first furrow with the embryonic axis was observed in three out of ten cases (Fig. IV).

The third experiment was made during the summer of '98. A broken bottle from the eel pond was brought into the laboratory containing a toadfish and a few eggs which had been already deposited by the fish.

On the morning of July 6 the fish was opened and the mature eggs allowed to flow out of the ovaries and fasten themselves on pieces of glass.

About one-third of the eggs set free from the ovaries became fixed to three glass plates. The rest failed to adhere and so were of no use in the experiment. Two minutes was sufficient time to allow for the eggs to become fixed. The plates were then inverted and placed on supports in a large shallow dish with sufficient sea-water to cover them.

The eggs were then fertilized, and soon after the water was siphoned off by means of a rubber tube and fresh sea-water

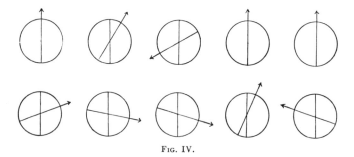

Fig. IV.

quickly supplied. The dish was so placed that the eggs could be observed from below.

At 4.30 P.M.—eight hours after fertilization—the first line of cleavage was noted in fourteen eggs, and the direction of the

furrow indicated on the paper oriented as in the earlier experiments.

On July 12, six days afterward, during which time the dish
containing the eggs was not moved, the axis of the embryo was
observed, and its direction indicated as before.   The result is

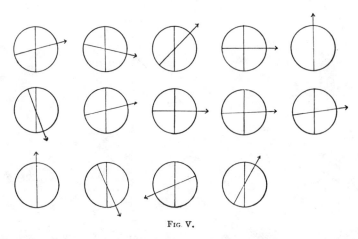

FIG. V.

seen in Fig. V.   Of the fourteen eggs only *two* show coincidence of the first cleavage furrow with the embryonic axis
(Fig. V, 5, 11).

Similar experiments with the eggs of *Amia* have been made
by Eycleshymer, '96.[1]   Observations were made on three sets
of eggs and the results are given in Fig. VI.

Twenty eggs developed normal embryos, and *two* showed
coincidence of the first plane of cleavage with the axis of the
embryo (Fig. VI, 2 and 6).   The following is quoted: "The
egg is oval in profile view, measuring in its longest diameter,
including the membrane, 2.5 to 3 mm.; in its shortest, 2 to
2.5 mm.   The freshly deposited egg is firmly fixed to the
object with which it first comes in contact by means of the
threads of the villous layer, which are elongated over the lower
hemisphere of the egg-membrane.

The eggs still attached to blades of grass or rootlets are
placed in shallow watch-glasses and held in position by weighting with small pebbles.   The watch-glasses are then placed

[1] *Journ. of Morph.*, vol. xii, p. 344.

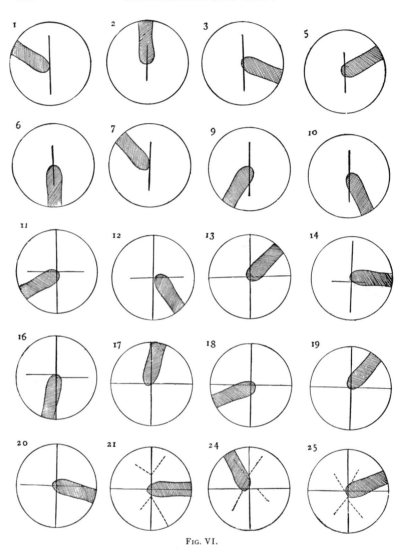

Fig. VI.

upon a mirror fastened to the stage of a dissecting microscope. We could thus observe the changes occurring on opposite sides of the egg without disturbing its position.

The elongated form of the egg of *Amia*, in a closely applied envelope, prevents rotation about its minor axes. It is, therefore, a favorable egg for ascertaining what effects, if any, gravity may have on the direction of cleavage, and for determining the relation of the early cleavage planes to the median plane of the embryo. Dean found " in the early stages of segmentation that the cleavage planes occurred in the normal way when the position of the egg was reversed." This is true not only for inverted eggs, but for eggs placed in any position whatsoever. It seems to follow that gravity can have no directing influence on the cleavage. In order to ascertain whether there is any constant relation of the embryonic axis to either of the first two cleavage planes, eggs were fixed in given position by weighting, as before mentioned, and a sketch of the early grooves was carefully made in each case. These grooves are easily identified for a long time in the lower hemisphere of the egg, even as late, in some cases, as the early stages of gastrulation. As the sketches made at successive intervals showed no movement of the egg during all this time, it seemed probable that the position of the egg remained practically unchanged up to the time when the median plane of the embryo was ascertainable. In some cases accidental markings on the surface of the egg remained in a fixed position until the embryo was well defined.

In 1893 Morgan [1] experimented with the eggs of teleosts. The pelagic eggs of *Ctenolabrus* and *Serranus* were selected. Morgan says : " My results show that there is no relation whatsoever between the cleavage planes of the egg and the median plane of the adult body. I have definite records of twenty-two eggs carefully marked. The results, expressed as nearly as possible, show for *Ctenolabrus* the plane of first cleavage agrees approximately with the median plane of the body in *five* cases, the plane of second cleavage with the median plane in *ten* cases. The median plane of the adult body lay between the first and second cleavage planes in *two* cases. For *Serranus* the first cleavage agreed in one case, the second agreed in two cases, and neither in two cases."

Two methods were employed. In one case the egg was watched continuously, and the median plane of the embryo corresponded to a plane midway between the first and second cleavage planes.

[1] *An. Anz.*, viii. Jahrg., p. 803.

The second method consisted in marking the membrane above the blastoderm and using such marks as points of orientation. "The success of the second method depended upon the close adherence between the egg and its membrane. After a very careful test it was found that even after rough treatment the egg retained a fixed position relatively to the surface markings. And again by watching individual eggs for some hours it was seen that the egg did not rotate within the membrane in the early stages or change its orientation with respect to the marks on the egg-membrane. In order to mark the eggs they were removed from the water and partially dried. A needle covered with finely divided carmine was drawn horizontally over the eggs. Small particles of carmine stuck to the membrane in many cases. The eggs were returned to the water and the best marked chosen."

Jordan and Eycleshymer[1] published a joint paper in 1893, from which it becomes plain that in the case of two species of *Urodeles* and two *Anura* there exists no constant relation between the early cleavage planes and the adult axes. "The first and second cleavage planes undergo extensive torsion, and the cells originally on one side of the mid-line come to lie on the opposite side."

Jordan[2] states, in regard to the cleavage in the newt, that "the total absence of any regularity in the arrangement of the cells is the most conspicuous feature."

Experiments of a somewhat different nature by Driesch, Wilson, and others have also shown conclusively that the first line of cleavage cannot have the significance which the earlier experiments of Roux would indicate. Recent studies in cell-lineage, together with the work of experimental embryologists, have thrown a great deal of light on the subject here considered, but without furnishing a decisive answer as to the meaning of cleavage.

The opinion expressed by van Beneden,[3] as early as 1883, that all bilateral animals would be found to agree in having the

---

[1] *Journ. of Morph.*, vol. ix, p. 407.
[2] *Journ. of Morph.*, vol. viii, p. 269.
[3] *Archives de Biologie*, IV, 1883.

axis of the body located and pre-marked by the first plane of cleavage, has steadily lost ground.

The study of cleavage has not yet given us any such fundamental law of development as the mosaic theory claims. The number of animals whose development does not conform to the supposed law of coincidence has increased so rapidly of late, that the only reasonable conclusion seems to be that while the first cleavage plane may coincide with the median axis of the embryo, as Roux and others have shown, it is not a constant rule in any single case, much less a universal law.

The opinion is gaining ground that the phenomena of cleavage are to be regarded as the expression of non-differential rather than qualitative divisions of the germinal material. The pressure experiments have proved that there can be extreme variation in cleavage produced without in any way interfering with the normal development of the embryo.

NOTE. — Batrachus tau = Opsanus tau. See *Bulletin of U. S. National Museum*. No. 47, Part 3, p. 2313. 1898.

6

WILLIAM MORTON WHEELER
1865–1937

Wheeler and Whitman met while Wheeler was serving as custodian to the Milwaukee Public Museum (1887–1890) and Whitman was directing the Allis Lake Laboratory nearby. In 1890, Wheeler followed Whitman to Clark University, where he received a fellowship and then a Ph.D. in 1892. He became an instructor at the MBL in 1891. After a year in Europe, he again followed Whitman, this time to the University of Chicago as instructor in embryology. Whitman worked hard to keep Wheeler at Chicago and wrote to President Harper several times of Wheeler's outstanding reputation and his excellence as a scientist. Unfortunately, Chicago did not provide sufficient support, and Wheeler left for the University of Texas in 1899 and thereafter to the American Museum and on to Harvard. While at Clark and Chicago, Wheeler focused on problems of embryology and morphology in marine life, but he quickly shifted to ants and other social insects, with a general interest in such problems as the possibility of emergent evolution. Whitman lamented to Harper in the 1890s that Wheeler and others were being pulled away from marine work by Chicago's demands for younger faculty to teach summer school. If Whitman could have had his way, Wheeler would have attended many more MBL sessions than he in fact did.

Unusually clear-headed, Wheeler provided lucid discussions of issues which often clarified the grounds of current debates. The essay reprinted here serves that purpose for a set of issues central to the MBL in the 1890s. This essay and others, such as the marvelously irreverant "The Dry-Rot of Academic Biology," also demonstrate Wheeler's classical training and his witty, smooth style of writing. By suggesting that people fall into two categories, the Heracliteans and the Parmenideans, he provided a working set of distinctions which others could then refine, reject, or endorse and thus advance toward understanding why people held the varying theories they did.

WILLIAM MORTON WHEELER.

CASPAR FRIEDRICH WOLFF AND THE *THEORIA GENERATIONIS*

WILLIAM MORTON WHEELER

Mag's die Welt zur Seite weisen,
Edle Schüler werden 's preisen,
Die an deinem Sinn entbrannt,
Wenn die Vielen dich verkannt.

GOETHE, *Morphologie*, p. 256.

THE universe which we apprehend—reducible in last analysis to various sequences and coexistences in time and space—seems to have a twofold aspect to the contemplative mind. The minds of some men are vividly affected by the *succession* of phenomena, the ceaseless current of events, the changes that alter the complexion of the world, the great qualitative and quantitative differences produced by these changes in that which we call matter. These observers may note the rhythm that is forever recurring in nature, the alternate repetition of day and night, the return of the seasons, the cyclical recurrence of stages in the development of living organisms—in short, the regular emergence from time to time of typical forms and conditions from the flowing current of events. This rhythm and repetition does not, however, produce the same deep impression on these observers as the successive and multiform changes themselves.

The other class of observer, although he may note the onrushing current of events, is more vividly impressed with the similarity of the forms and conditions that recur from time to time and from place to place. The attention is fixed on these recurring objects and conditions, and gradually builds them into general concepts that ultimately acquire a stability which nothing can shake. The movement of the stream of phenomena takes a subordinate position in consciousness, and the mental activities attach themselves by preference to stable, island-like forms and principles.

Thinkers from the earliest times to the present day seem to be referable to one or the other of these two classes. The differentiation begins in early Greek philosophy with men like Heraclitus and Parmenides. To Heraclitus the world was an unceasing flux—παντα ῥει, ουδεν μενει, *all things are flowing, nothing is standing still.* All things are forever *becoming*, nothing ever *is*. Parmenides, who fixed the trend of the Eleatic school, belonged to the other class. He is the philosopher of rest. The chaotic, multiform world of Heraclitus, forever in motion, becomes for him merely a world of nonexistent appearances, a shifting phantasmagoria, and only being *is*—the *absolute*—the *one*, forever at rest.

The contrast in these two views reappears between Aristotle and Plato. This difference is seen in the all-pervading movement as conceived by Aristotle in his Physics, in contrast with the "ideas" of Plato. Movement to Aristotle is "something very analogous to our modern biological conception of transformation in development, for he analyzes 'movement' as every change, as every realization of what is possible." [1] Plato, on the other hand, under the influence of Parmenides and the philosophy of rest, emphasizes the forms and qualities that keep recurring to our minds in time and space, generalizing them into his "ideas" and endowing them with all the attributes of reality. [2] He would say, *e.g.*, of a living animal as it stands before us: "This animal as we see it does not exist in reality, but is only an apparition, a continual becoming, a relative existence, which can as well be called nonexistent as existent. The idea alone actually exists which is represented in this animal, or the animal itself (αυτο το θηριον). This idea is independent of everything; it exists by itself; it has not become; it does not decay, but exists always in the same

[1] Osborn, H. F. From the Greeks to Darwin. New York, Macmillan & Co., 1894. p. 50.

[2] See Pater, Walter. Plato and Platonism, Chaps. I, II.

manner (ἀεὶ ὂν, καὶ μηδέποτε οὔτε γιγνόμενον, οὔτε ἀπολλύμενον). If we can recognize the *idea* in this animal, it is immaterial and unimportant whether we are looking at the animal now before us or its ancestor that lived a thousand years ago, or whether the animal is here or in a distant land, or whether it appears in this or that manner, position or action, whether, finally, it be this or another individual of the same species : all this is unessential and appertains only to appearances : the *idea* of the animal alone really *is*, and really is an object of the understanding." [1]

It would not be difficult to trace the Heraclitean conception of the flux through Aristotle down to such modern philosophers as Hegel and Herbert Spencer, and to trace the Platonic idea, through the λογος σπερματικος of the Stoics, the *forma substantialis* and the *causae primordiales* of the scholastics, to Kant's Ding-an-sich, Schelling's Absolute, and the Platonic idea as adopted by Schopenhauer. But the tracing of these conceptions in detail would lead us far afield in metaphysics. I should beg your indulgence for mentioning these matters did they not seem to me to be, in some measure, necessary to a proper understanding of the two great views of embryonic development that have been and still are held by thinking students of nature — *preformation* and *epigenesis*.

The development of the living organism is the most striking special case of development we know. The development of what appears to be a simple egg, within a comparatively short time, and beneath our very eyes, into a complex living animal, is development *par excellence* — the very perfection of that development which is more dimly apprehended in the much slower growth of worlds, of human societies and human institutions. Hence we do not wonder that the development of the individual organism has become one of the main tests of two alternative views which, with a more general application, have from the earliest times vexed philosophic thinkers.

Under the influence of the Christian church the Platonic conception seems to have led to the notion of the special crea-

[1] Schopenhauer, A. Die Welt als Wille und Vorstellung. Leipzig, Brockhaus, 1888. Bd. i, p. 203.

tion of fixed types or forms. It culminated in that finished theory of predelineation in embryonic development known as *emboîtement.*[1] This was, in reality, the very negation of all development, since the theory held that all the individuals of a species had been created simultaneously for all time.[2] In the forcible language of the last century, Eve's ovary contained the compressed and diminutive germs of all coming human beings incapsulated one within the other. Such a theory could arise only from overestimation of the definitive form attained through development, and an underestimation of the changes undergone by the egg during its development. The typical adult form usurped the theorist's attention, and the elaborate process whereby the type was gradually realized shrunk to a mere unshelling and subsequent growth in size of the next individual in order in the incapsulated series.

For the theory of *emboîtement* the creation not only of every species, but of every individual organism on our planet, by a single preadamite fiat, was a necessary postulate. The rival theory, epigenesis, implied in the cosmology of Heraclitus and easily traceable to Aristotle, starts with a simple form of unorganized matter, which through the agency of certain forces undergoes the complicated changes that finally result in the adult living organism. The homogeneous becomes the heterogeneous. The creation of new organisms is no longer conceived as having taken place once for all in a remote and inscrutable past, but as taking place everywhere and at all times. An exaggeration of epigenesis is spontaneous genera-

[1] Passages which show the close genetic relationship of Neo-Platonic and Christian thought on the subject of creation are not infrequent in the writings of the Church Fathers. The following quotations from Augustine clearly express the idea of *emboîtement:* " Sicut autem in ipso grano invisibiliter erant omnia simul, quæ per tempora in arborem surgerent, ita ipse mundus cogitandus est, cum Deus simul omnia creavit, habuisse simul omnia, quæ in illo et cum illo facta sunt, quando factus est dies : non solum coelum cum sole et luna et sideribus . . . sed etiam illa quæ aqua et terra produxit, potentialiter atque causaliter priusquam per temporum moras ita exorentur, quomodo nobis jam nota sunt in eis operibus, quæ Deus usque nunc operatur." De Genesi ad lit., v, 45. "Omnium quippe rerum quæ corporaliter visibiliterque nascuntur, occulta quædam semina in istis corporis mundi hujus elementis latent." De Trinitate, iii, 8.

[2] " Qui igitur systemata praedelineationis tradunt, generationem non explicant, sed, eam non dari, affirmant." C. F. Wolff, Theoria Generationis, 1759, p. 5.

tion.   Aristotle even believed that mud could become earth-worms and earthworms become eels.[1]

Before the end of the past century these two views of devel-opment which I have attempted to trace back respectively to Aristotle and Plato had assumed definite and contrasting forms.   Bonnet, Haller, and Leibnitz, following a Platonizing tendency in dealing with natural phenomena, had elaborated and accepted the theory of *emboîtement*, or "evolution," as the word was then understood.   Bonnet's contributions to this view have been adequately presented by Professor Whitman, in his lectures to the members of the Marine Biological Laboratory during the summer of 1894.[2]   Haller, justly styled by his con-temporaries an "abyss of learning," though devoted to *emboîte-ment*, had too great a store of mental riches to give himself up year after year, like Bonnet, to exhaustive rumination on a single theory.   The opinion of Leibnitz on *emboîtement* is not so generally known, and may be considered briefly.   The philosopher of a preëstablished harmony could hardly overlook a theory like that of predelineation.   Like many philosophers of the present day, Leibnitz was glad to accept the theories of contemporary scientists, weave them into his general scheme, and, without adding anything really new, again present them to the public, heavier with the weight of his name and authority. In his "Monadologie," he says[3]: "Philosophers have had much difficulty in dealing with the origin of forms, entelechies, and souls.   Of late, however, careful investigations on plants, insects and animals, have led to the conclusion that in nature organic bodies never arise from chaos or decomposing matter, but always from germs (semences), in which, *without a doubt*, they are already preformed.   Hence we may conclude that in this Anlage not only do organic bodies exist before generation, but that there is a soul in these bodies, in a word, the indi-

[1] Aristoteles. Ἰστοριαι περι ζωων. Ed. Aubert u. Wimmer. Leipzig, 1868. ii, 6. 16. pp. 56 and 58. — J. Bona Meyer. Aristoteles Thierkunde. Berlin, 1855. pp. 97, 98.

[2] Whitman, C.O. (1) "Bonnet's Theory of Evolution a System of Negations." (2) "The Palingenesis and the Germ Doctrine of Bonnet," *Biological Lectures* (1894). Boston, Ginn & Co., 1895.

[3] Leibnitz, Op. Phil., p. 711.

vidual itself, and that reproduction is merely a means of enabling this individual to undergo a greater change in form, to become an individual of a different kind. Something similar to generation is seen when maggots become flies and caterpillars butterflies." At another place, in the "Theodicee," he says,[1] after referring to the microscopic observations of Leeuwenhoek: "Thus I would contend, that the souls, which are some day to become human souls, were already present in the germ like the souls of other species, that they have always existed in our fore-fathers as far back as Adam, *i.e.*, since the beginning of things, in the form of organized bodies." These remarks of Leibnitz are the *ne plus ultra* formulation of the theory of *emboîtement* — its extension to embrace not only the physical but also the psychical and spiritual aspect of living things.

It is, perhaps, easy to understand how philosophical and religious preconceptions could give this final form to the theory of *emboîtement*. Other considerations, however, of a more real and scientific character seem to have led men's minds in the same directions. The microscope, invented in the sixteenth and bequeathed to the seventeenth century, had profoundly influenced speculation. Magnification had revealed, as if by magic, the existence of a great world of structures undreamed of by the greatest intellects the race had hitherto produced. The authority of the ancients weakened perceptibly, for little value could thenceforth be attached to their opinions on the nature of the great world that stretched out beyond the confines of unaided vision. The mind, full of the great microscopic discoveries of the time, was carried away by its own inertia, and, outrunning the instrument, first dreamed of and then believed in the existence of structures too minute to be revealed by the available lenses. This speculation was, perhaps, justifiable, except when it undertook to define the precise nature of what was at that time an ultra-microscopic region. It was natural but erroneous to conceive unseen structures as diminutive duplicates of the seen. The verisimilitude of this error increased when it became apparent that the microscope was unable to resolve perfectly transparent structures even of

[1] Op. Phil., p. 527.

considerable size. And here the theorist triumphed over the empirical observer, for he could assert, what was not easily disproved, that owing to their transparency the microscope must ever fail to reveal the germs incased one within the other.

The Siegfried destined to overcome this monstrous theory of *emboîtement*, a theory not only false in itself, but one jealously guarding the problem of development, and preventing all access to it, as the dragon guarded the treasure of the Niebelungen, was *Caspar Friedrich Wolff*. Wolff, one of the many great intellects that northern Germany has produced, was born in Berlin in 1733. You will find nearly all that is known of his life in a letter by his amanuensis Mursinna to Goethe, published by the great German poet in his *Morphologie*.[1] The scant facts of this letter, with Wolff's own writings, in which his personality is studiously kept in reserve after the manner of scientific men, leaves us with a sense of uncertainty not entirely free from sadness. We long to know more of this sweet-natured student who, at the early age of six-and-twenty, was an intellectual giant, defending an epoch-making thesis, the *theoria generationis*, simply *pro gradu doctoris medicinae*.

Before giving a brief account of this *Theoria* it may be well to try to form some idea of its author's general mental characteristics. Wolff was a disciple of Aristotle. The training of the schoolman is only too apparent in all his scientific writings, apart from Mursinna's statement[2] to the effect that when Wolff was lecturing on medicine in Berlin " he taught logic probably better than it had ever been taught before, and applied it in particular to medicine, thereby creating, so to speak, a new spirit in his hearers, so that they were enabled to understand and assimilate his other teachings more easily." His skill in deductive logic seems to have been noticed by Sachs,[3] who claims that some of Wolff's observations on plant structure " are highly inexact, and influenced by preconceived opinions, and his account of them is rendered obscure and often quite

---

[1] Goethe. Morphologie, 1820, pp. 252–256.
[2] Goethe. Morphologie, p. 254.
[3] Sachs. History of Botany, p. 251.

intolerable by his eagerness to give an immediate philosophic explanation of objects which he had only imperfectly examined." The same statement may be extended to many of his zoölogical observations, but this is far from convincing us that his method of investigation was at fault. Wolff's method, which did not differ from that of the scientist of to-day, was, if anything, more admirable than his observations. The very fact that he was full of his Aristotelean hypothesis of epigenesis places him head and shoulders above the investigators both of his day and of to-day, who naïvely believe that they are starting their investigations on a solid foundation of facts divorced from all theory. Even Sachs admits that Wolff's phytotomical work, though poor from the standpoint of observation, was the most important that appeared in the period between Grew (1682) and Mirbel (1802), " because its author was able to make some use of what he saw, and to found a theory upon it." [1]

Apart from this preconceived hypothesis of epigenesis it is surprising with what perfect *naïveté* Wolff approaches the phenomena to be observed. Armed with his microscope, which it does not require a Sachs to tell us "was of insufficient power and its definition imperfect," he entered what was practically an unknown domain, peopled only with the figments of the predelineationists. The fascination of the growing plant and developing embryo soon possessed him and never afterwards left him. During his long life he returned again and again to the study of the chick. Those who teach embryology year after year cannot fail to appreciate Wolff's power and greatness when they observe the superficial impression left on nine-tenths of the students who study the developing chick in the well-equipped laboratories of to-day.

Wolff's instruments, poor as they were, enabled him, nevertheless, to traverse a considerable and very significant portion of the region that lies beyond the boundary of our unaided vision. What he saw there at once convinced him that *emboîtement* was a myth. We should expect so young a man as Wolff was when he wrote the *Theoria* to do two things — to repeat his main thesis *ad nauseam*, and to be rather unsparing

[1] Sachs. History of Botany, p. 251.

of his opponents. He did neither. His main contention is clear enough, although not often expressly stated. He rarely refers directly to the theory of predelineation and when he does there is no sting in his refutation.

Stripped of many details that are somewhat wearisome to the modern reader, the result of Wolff's observations may be expressed in his own words taken from the very middle of the *Theoria*.[1] I translate : " In general we cannot say that what cannot be perceived by the senses does not therefore exist. This principle is more facetious than true when applied to these observations. The particles which constitute all animal organs in their earliest inception are little globules, which may always be distinguished under a microscope of moderate magnification. How, then, can it be maintained that a body is invisible because it is too small, when the *parts* of which it is composed are easily distinguishable ? " If we of to-day read in the place of "globules" the word "cells," which are what Wolff actually saw and distinguished in both plants and animals, we shall have no difficulty in understanding how his observations disproved *emboîtement*, at least for any one who would take the trouble to repeat them. Wolff had looked further than the adult form and had found not a series of similar, incapsulated embryos, but a single embryo made up of a vast number of minute particles, the cells, closely resembling one another, but placed side by side. There was no expanding of a preëxisting organism till it entered the field of vision, but a host of minute and always visible elements that assimilated food, grew and multiplied, and thus gradually in associated masses produced the stem, leaves, stamens — in short, every organ of the plant. This he shows in the first part of the *Theoria*. In the second part, carrying on the same method, he shows how in the animal body the heart, blood vessels, limbs, alimentary canal, kidneys, etc., arise in a similar manner.

The third part of the work is devoted to theoretical considerations. Wolff conceives living things to be constructed like a plant, to consist of a main stem, or trunk with roots and branches. In the embryo of the bird the umbilical duct cor-

[1] Theoria Generationis, p. 72.

responds to the stem of the plant; the blood vessels of the vascular area that bring the nutriment from the yolk to the embryo are the roots ; the organs and appendages of the embryo correspond to the branches of the plant.[1] The organism starts out on its development with a stem which is to connect it with the source of nutrition on the one hand and its branches on the other. All the substance of the embryo is originally unorganized, inorganic. Organization sets in at one point in the stem and thence gradually spreads to the tips of the branches. A branch is first formed as a little bud of unorganized substance, then the sap (in plants) or the blood (in animals) flows into it from the adjacent organized part ; thus it becomes organized, and the process continues till the organism has acquired its definitive size and development. The blood or sap is propelled into the unorganized substance, consisting of globules, by a peculiar force — Wolff's *vis essentialis,* which is defined in the opening chapter of the *Theoria,* and was made the special subject of Wolff's last work, written thirty years later.[2] Organization of the unorganized substance is the combined result of this *vis essentialis* and a property which Wolff calls *solidescibilitas,* a tendency to solidify, shown most clearly in the formation of the walls of plant cells. The *vis essentialis* propels liquid nutriment into the dense unorganized matter already present. The paths along which it flows become the cavities of the blood vessels or plant vessels not before existent. The liquid nutriment solidifies to form more unorganized substance, by intussusception with that already present, and the part grows. Wolff explains the origin of the kidney which he discovered in the chick — the Wolffian body, or mesonephros, as we now call it — in a similar manner. Here it is the urine that is impelled by the *vis essentialis* into a mass of preëxist-

---

[1] Wolff (Theorie von der Generation, 1764) compares four-footed animals to pinnatifid leaves and "the bat is a perfect leaf — a startling statement, but, as I have shown, the analogy is not chimerical, for the *mode of origin of the two is the same.*" Quoted by Huxley ("The Cell Theory," *Brit. and Foreign Medico-Chir. Review,* vol. xii, 1853).

[2] Wolff, C. F. Von der eigenthümlichen und wesentlichen Kraft der vegetabilischen sowohl als auch der animalischen Substanz. St. Petersburg, 1789.

ing, unorganized substance, and the paths along which it flows become the lumina of the uriniferous tubules and the ureter.

We know that Wolff's main error lay in grossly underestimating the complexity of the problem he attempted to solve. This has always been a great pitfall in attempting an explanation of life. Perhaps it is well that it is so, for Wolff would hardly have had the heart to attempt it if he could have seen the problem with our eyes. And may not we, too, daily commit the same blunder when we lend a willing ear to those who regard living protoplasm as nothing more than a "complex chemical compound"?

Wolff accepted a simple substance as the basis of life because he was unable to detect structure in the embryo beyond a certain limit which happened to coincide with the limits of magnification of his lenses. We should suppose that Wolff would have longed for a better lens and have at least suspected the possible existence of some kind of structure beyond that which he could detect. Instead of doing this, however, he writes the following remarkable sentences which will draw a smile from the modern searchers after centrosomes [1]: "No one has ever yet, with the aid of a stronger lens, detected parts, which he could not perceive by means of a weaker magnification. These parts either have not been seen at all, or they have appeared of sufficient size. That parts may remain concealed on account of their infinitely small size and then gradually emerge, is a fable." There it is in cold Latin! Was Wolff merely nodding when he wrote this, or was he trying to hoodwink the predelineationists into believing that he had seen everything that was worth seeing in the embryo?

In the closing paragraph of his great work on the development of the intestinal tract, a work which appeared in 1768, some nine years after the *Theoria*, Wolff seems to rise to a clearer perception of the complexity of the problem. He appears to be far more doubtful concerning the way in which simple matter becomes organized. Referring to the development of the anterior body wall, he says [2]: "This is one of the

[1] Theoria, Sect. 166.
[2] Wolff, C. F. Ueber die Bildung des Darmkanals im bebrüteten Hühnchen. Uebersetzt von J. F. Meckel. Halle, 1812, p. 245.

most important proofs of epigenesis. We may conclude from it that the organs of the body have not always existed, but have been formed successively: no matter how this formation has been brought about. I do not say that it has been brought about by a combination of particles, by a kind of fermentation, through mechanical causes, through the activity of the soul, but only that it has been brought about."

Remaining within the province of observation which he staked out for himself, and pursuing his excellent method, Wolff was not only able to undermine the theoretical edifice of the predelineationists, but also to lay the foundations for future structures of great promise. Thus all conscientious investigation with good methods leads to subordinate facts of value besides the main line of facts accumulated in support of the theory in hand. Wolff was a biologist in the true sense of the word. He regarded plant and animal life as but slightly different aspects of a single set of phenomena. It can be shown that he anticipated to some extent the modern theories of protoplasm and the cell.[1] According to Sachs "it was Wolff's doctrine of the formation of cellular structure in plants which was in the main adopted by Mirbel at the beginning of the present century," and "the opposition which it encountered contributed essentially to the further advance of phytotomy." [2]

The theory of the metamorphosis of plants, usually attributed to Goethe, was clearly expressed by Wolff. In fact, Wolff seems to have had clearer notions on the subject than Goethe, according to Schleiden's statement.

To embryology Wolff made many valuable contributions, not the least of which was his description of the formation of the intestinal tract of the chick. This work was styled by Carl Ernst von Baer "die grösste Meisterarbeit, die wir aus dem Felde der beobachtenden Naturwissenschaften kennen." It was published in Latin in the twelfth and thirteenth volumes of the *St. Petersburg Commentaries*, where it lay buried and forgotten till it was unearthed and translated into German by

---

[1] Cf. Huxley.

[2] Sachs. Hist. of Botany, p. 250. For the relations of Wolff's views to those of Schleiden and Schwarm, see Huxley, The Cell Theory.

the younger Meckel in 1812 and used for the purpose of refuting some of Oken's erroneous views on the development of the alimentary tract.

In general it may be said that the effect of Wolff's work on his contemporaries was anything but immediate.[1] There are writers who even doubt the truth of the oft-repeated statement that Wolff refuted the theory of predelineation. Sachs, *e.g.*, speaking of Wolff's *Theoria*, says that the "weight of his arguments was not great" and that "the hybridization in plants which was discovered at about the same time by Koelreuter supplied much more convincing proof against every form of evolution."[2] We cannot lay much stress on this statement, which seems to imply, what some physiologists seem never to tire of implying, that evidence derived from experiment is *eo ipso* more convincing than evidence derived from observation. It is certain that the predelineationists had considered the case of hybrids, for did not the ever-watchful Bonnet endeavor to explain the origin of the mule on the assumption of *emboîtement?* And why should Koelreuter's plant hybrids have more value in refuting *emboîtement* than that commonest of all hybrids, the mule? If Sachs wishes to imply that at the present day we should regard the evidence from hybrids as a complete and satisfactory refutation of the theory of *emboîtement*, we may assent; but this is not tantamount to saying that in the latter half of the eighteenth century it was Koelreuter and not Wolff who refuted the theory of evolution. Perhaps it would be better to leave this question of the relative merits of Wolff and Koelreuter to the student who has the time and the opportunity to study all the relevant literature of the closing decades of the eighteenth century.

Wolff's position in the history of thought on the subject of organic development becomes somewhat clearer when we compare him with Darwin, for whose coming he helped to prepare

---

[1] "Though every reader of the Theoria Generationis must see that Wolff triumphantly establishes his position, yet, seventy years afterwards, we find even Cuvier (Histoire des Sciences Naturelles) still accrediting the doctrine of his opponents." — Huxley, The Cell Theory.

[2] Sachs. History of Botany, p. 405.

men's minds. Wolff's *Theoria* was published in 1759; Darwin's *Origin of Species* in 1859. Wolff had been preceded by Harvey in much the same way as Darwin was preceded by Lamarck. Both Wolff and Darwin were ideal investigators, patterns for all time. Darwin's love of truth, his perfect fairness and modesty withal, seem to have been Wolff's possession also. This is shown in a letter to Haller,[1] thanking the great champion of *emboîtement* for his kindly notice of the *Theoria* in his *"Elementa": "*I thank you for wishing me well, for loving me, sublime man, although you have never seen me, and know me and my character only from my letters. May God reward you for this, since I can never hope in all my life to attain to such distinction, that I may show you worthy acknowledgment of your goodness, if you will not receive in lieu of it my everlasting veneration of your intellect. And as to the matter of contention between us, I think thus: For me, no more than for you, glorious man, is truth of the very greatest concern. Whether it chance that organic bodies emerge from an invisible into a visible condition, or form themselves out of the air, there is no reason why I should wish that the one were truer than the other, or wish the one and not the other. And this is your view, also, glorious man. We are investigating for truth only; *we seek that which is true.* Why, then, should I contend with you? Why should I withstand you, when you are pressing towards the same goal as myself? I would rather confide my epigenesis to your protection, for you to defend and elaborate, if it is true; but if it is false, it shall be a detestable monster to me also. I will admire evolution, if it is true, and worship the adorable Author of Nature as a divinity past human comprehension; but if it is false, you, too, even if I remain silent, will cast it from you without hesitation."

Both Wolff and Darwin devoted their lives to the investigation of the same great problem — the development of life on

---

[1] Epistolae ad Hallerum, October, 1766. Quoted from Alf. Kirchhoff, "Caspar Friedrich Wolff. Sein Leben und seine Bedeutung für die Lehre von der organischen Entwickelung," *Jen. Zeit. f. Med. u. Naturwiss.*, Bd. iv, Heft 1, 1868, pp. 193–220. This valuable essay has been of great assistance in the preparation of my lecture.

our planet. Both found answers to their respective parts of this problem. Wolff published his answer when he was very young; Darwin waited till he was well along in years. Each was confronted by a formidable, clearly formulated theory of special creation. The theory that confronted Wolff was special creation of all *individual* organisms by a preadamite fiat. Darwin was confronted by a theory of the special creation of all .the *species* at the same inscrutable time. This view had found favor with such eminent men as Linné, Cuvier, L. Agassiz, Owen, and the numerous systematists who followed in their footsteps, making species Platonic ideas, just as individual organisms had been made Platonic ideas in Wolff's day. During the closing half of the eighteenth century it became clear to thinking men that individual organisms always have an epigenetic origin from preëxisting individuals. The closing half of the present century has been consumed in demonstrating that species always arise from preëxisting species.

Both Wolff and Darwin collided with prevailing theological views. Darwin's experience in this matter is well known, and perhaps the less said about it the better. It is not so generally known that Wolff's failure to establish himself as a professor in Germany, and his departure in 1769 for St. Petersburg, where he spent the remainder of his life, was probably due not only to professional jealousy, but also to a certain antagonism on the part of religious contemporaries. In his letters Haller often warned Wolff of the dangers of his views to religious dogma, and endeavored to persuade him to abandon them "on grounds of utility." [1] Just before leaving for St. Petersburg Wolff wrote the following to Haller: "There is, of course, no reason why a divine being should not exist, even if organic bodies are formed by natural forces and through natural causes; for these very forces and causes, yes, Nature herself, has as much need of an Originator as organic bodies; still the evidence would be far more cogent and apparent, if we should find from contemplation of Nature that her individual products, the organic bodies, required a Creator, and that nothing organic could be produced through natural causes." Does not this

[1] Kirchhoff, A., *l.c.*

remind us of the following passage in the last chapter of Darwin's *Descent of Man?* " I am aware that the conclusions arrived at in this work will be denounced by some as highly irreligious ; but he who denounces them is bound to shew why it is more irreligious to explain the origin of man as a distinct species by descent from some lower form, through the laws of variation and natural selection, than to explain the birth of the individual through the laws of ordinary reproduction. The birth both of the species and of the individual are equally parts of that grand sequence of events, which our minds refuse to accept as the result of blind chance. The understanding revolts at such a conclusion, whether or not we are able to believe that every slight variation of structure, — the union of each pair in marriage, — the dissemination of each seed, — and other such events, have all been ordained for some special purpose."

Both Wolff and Darwin left their theories unfinished. They maintained a transformation of simpler into more complex matter, but they did not succeed in demonstrating how this transformation is accomplished. I have already quoted Wolff's confession of ignorance of the way in which epigenetic development is brought about. The doubts entertained by Darwin and his successors concerning the adequacy of natural selection as a complete explanation of descent are familiar to us all. The absolute completeness of the old *emboîtement* and special creation hypotheses doomed them to a speedy death. Wolff's and Darwin's hypotheses have lived because they represented only parts of a great truth. On this account, also, they have supplied and will continue to supply powerful incentives to investigation.

Both the theory of epigenesis and the modern theory of descent are manifestly imbued with the old Heraclitean and Aristotelean conception of heterogeneity arising from homogeneity in a continual flux of events. We have come to regard this as the essence of evolution. We still repeat Herbert Spencer's definition[1] : " Evolution is an integration of

[1] Spencer, Herbert. First Principles. New York, Appleton & Co., 1886. p. 396.

matter and a concomitant dissipation of motion; during which the matter passes from an indefinite, incoherent homogeneity to a definite, coherent heterogeneity; and during which the retained motion undergoes a parallel transformation." If Wolff could have read this sonorous definition he would probably have accepted it as the expression of a general truth, as it is accepted to-day. Still when we compare our views on the development of living organisms with those of Wolff, we observe a vast difference, to which Professor Whitman calls attention when he says[1]: "The indubitable fact on which we now build is no bit of inorganic homogeneity, into which organization is to be sprung by a coagulating principle, or cooked in by a *calidum innatum*, or wrought out by a spinning archaeus, but the *ready-formed, living germ, with an organization cut directly from a preëxisting, parental organization of the same kind.*

"The essential thing here is, not simply continuity of germ-substance of the same chemico-physical constitution, but *actual identity of germ-organization with stirp-organization.*"

To-day we recognize three conditions of matter: dead matter, undifferentiated living protoplasm, and differentiated living protoplasm. The germ is cut from its parent as organized but undifferentiated living protoplasm; during ontogeny it is converted into differentiated living protoplasm. Wolff saw no difficulty in leaping with a bound from dead matter to highly differentiated living protoplasm. There are scientific acrobats still living who are not at all afraid of taking a like hop, skip, and jump. The majority of biologists, however, are too heavy with the past century's observations on living matter to be able to leap so fast and so far. They find the distance between dead matter and undifferentiated protoplasm enormous — a chasm so wide and deep that they prefer making a long detour to attempting a perilous leap across it. They are more intrepid in passing over the gap, great as it undoubtedly is, which separates the undifferentiated from the differentiated phase of organized matter, the germ from the adult. Current discussion is, therefore, mainly limited to the valuation of the

[1] Whitman, C. O., "Evolution and Epigenesis," *Biological Lectures* (1894). Boston, Ginn & Co., 1895. p. 212.

extent of complexity in the germ as compared with the complexity of the fully developed organism.

He who finds little difficulty in passing from the simple to the complex, from the homogeneous to the heterogeneous, will take an epigenetic view of development. The physiologist, who deals with processes, who is ever mindful of the Heraclitean flux, inclines naturally to this view. On the other hand, he who readily idealizes and schematizes, whose mind is endowed with a certain artistic keenness, an appetite for forms and structures, and a tendency to make these forms final patterns, eternal molds, more permanent than the substance that is poured into them — such a one will find more difficulty in understanding *how* the homogeneous can become the heterogeneous. Of this type is the modern morphologist who is continually diagrammatizing, who has his eye fixed on complex static structures and conceives the continually changing form of the developing egg as a series of kinematograph pictures in three dimensions of space. He is as much inclined to Platonize as is the modern physiologist to reason along lines suggested by Aristotle. He is by nature a preformationist.

Just as Wolff's followers have split into two schools — one believing in little, the other in much preformation in the germ — so Darwin's followers have split into two schools, the Neo-Lamarckians and the Neo-Darwinians, in obedience to the two psychological tendencies to which I have called your attention. The Neo-Darwinians, in laying great stress on the segregation, stable and complex *intrinsic* structure of the germ plasma and its importance as a vehicle of hereditary characters, and in attributing less value to the *extrinsic* factors, like food and environment, are allying themselves with theorists of the type of Parmenides and Plato. On the other hand, the Neo-Lamarckians who believe in the permanent change-producing effects of the *extrinsic* factors (environment, etc.) on structure, and attribute less value to the architecture of an *intrinsic* vehicle of heredity (germ plasma), range themselves with Heraclitus and Aristotle.

The amount of differentiation displayed during the ontogeny of an organism or during its phylogenetic history will be

differently estimated by different workers, till we are in possession of some means of mathematically measuring differentiation and variation. The demand for mathematical measurement
is already being made in certain quarters, and this demand
progressing science will undoubtedly supply. At present we
are quite adrift in our discussions, so long as we ignore the
more general and philosophical aspect of the question and
thereby overestimate its simplicity. Even if we accept differentiation and the interaction of differentiated products as the
root ideas of the evolution of the individual and of the race,
we are still at a loss to understand how the initial differentiation arose — how the homogeneous first became the heterogeneous. Lloyd Morgan has recently expressed this hopeless
and baffling search for initial differentiation as follows [1] : —

" Assuming, with the nebular hypothesis, a primitive fire-mist, we
must assume also an environment from which it is already differentiated and to which its heat energy is communicated by radiation.
Or if we accept the meteoric hypothesis, we must grant the existence
of already differentiated cosmic dust and the interaction of its constituent meteors. If we give yet freer rein to the speculative tendency, which, chastened or running riot, is man's blessing or curse,
and, straining our mental vision, search deeper still into the beginnings of our universe, to find in the homogeneous substance that
Sir William Crookes calls *protyle*, the stuff from which the chemical
elements were differentiated ; even in this dim and wholly hypothetical region we are forced to assume, as the antecedent conditions of
differentiation, transformations and redistributions of energy, implying a prior differentiation to render such interaction conceivable.
Or if, once more, we conceive the elemental atoms as vortex rings,
differentiated from the ether and thenceforth interacting, even here
at the very threshold of differentiation, we seek for an answer to the
question : Under what physical conditions did such vortex motion
originate ? "

[1] Morgan, Lloyd, " The Philosophy of Evolution," *Monist*, July, 1898,
p. 489. Weismann, too, expresses this difficulty in his "Germinal Selection ":
" Die sog. ' epigenetische ' Theorie mit *gleichen* Keimeseinheiten ist deshalb
eigentlich nichts Anderes, als eine Evolutionstheorie mit unbewusster Zurück-
verlegung der Anlagen in die Moleküle und Atome, eine, wie mir scheint, unstatt-
hafte Vorstellung. Eine *wirkliche* Epigenese aus völlig *gleichartigen*, nicht bloss
aus untereinander *gleichen* Einheiten ist nicht denkbar."

The pronounced " epigenesist " of to-day who postulates little or no predetermination in the germ must gird himself to perform Herculean labors in explaining how the complex heterogeneity of the adult organism can arise from chemical enzymes,[1] while the pronounced " preformationist " of to-day is bound to elucidate the elaborate morphological structure which he insists must be present in the germ.   Both tendencies will find their correctives in investigation.

[1] See, *e.g.*, Loeb, J., " Assimilation and Heredity," *Monist*, July, 1898, p. 555.

7

CHARLES OTIS WHITMAN
1842–1910

After attending Agassiz's Penikese Island school in 1873 and 1874, Whitman received his Ph.D. in Leipzig under Rudolf Leuckart for his work on the leech *Clepsine*. He taught at the Imperial University of Tokyo for two years, worked at Harvard's Museum of Comparative Zoology under Alexander Agassiz, and directed the Allis Lake Laboratory in Milwaukee. By 1890 he headed the biology department in the new Clark University for graduate research. Many of the people who lectured and taught at the MBL in the early 1890s were associates of Whitman's at Clark. Most of the others were acquaintances from the Museum of Comparative Zoology and the Boston area. In 1892 Whitman became head of the biology department at the new University of Chicago, taking most of Clark's biology department with him. He continued to advocate the central importance of marine study throughout the 1890s. Yet as Chicago repeatedly failed to provide a marine laboratory or the inland experimental station which Whitman urged, he established his own research facility in his backyard. There, Whitman set up a pigeon colony and began to study the life history, evolution, and behavior of those birds. Hauling the back and forth to Woods Hole for several summers proved expensive and tiring, but Whitman had committed himself to pigeons and so persisted. This dedication brought the study of instinct and behavior to the MBL in the late 1890s as it had not been before and has not since.

The list of lecturers and participants in the neurology seminars which Whitman inaugurated at the MBL provides a roster of leading American behaviorists of the day. Whitman saw many of the problems which others called psychological as being essentially biological, and fostered the development of comparative physiological psychology within biology at both the MBL and Chicago. This paper represents Whitman's classic summary of contemporary issues in animal behavior.

CHARLES OTIS WHITMAN.

ANIMAL BEHAVIOR

C. O. WHITMAN

"*Natura non facit saltum*, is applicable to instincts as well as to corporeal structure."
— DARWIN, *Origin of Species*, p. 231.

CONTENTS.

ANIMAL behavior, long an attractive theme with students of natural history, has in recent times become the centre of interest to investigators in the field of pyschogenesis. The study of habits, instincts, and intelligence in the lower animals was not for a long time considered to have any fundamental relation to the study of man's mental development. Biologists were left to cultivate the field alone, and psychologists only recently discovered how vast and essential were the interests to which their science could lay claim.

The contribution which I have to offer aims at no extensive exposition of the subject, but rather to call attention to some phenomena which I have observed, and to connect therewith such interpretations and theoretical considerations as may come within the sphere of general biology.

In animal life there are many interesting modes of keeping quiet, which are instinctive and adapted to special purposes. It is a very general means of concealment and escape from enemies. For illustration, we may take first the leeches, animals relatively low in the scale. One of the lower and least active forms, occurring everywhere in ponds, lakes, and streams, is known under the generic name *Clepsine*. There are many species, varying from one-quarter inch to one inch or more in length. They are found on their regular hosts, turtles, frogs, fishes, molluscs, etc., or on the under side of stones, boards, branches, or other submerged objects near the shore. One of the larger species, found often in large numbers on turtles, will be favorable for observation.

BEHAVIOR OF CLEPSINE.

a. *Deceptive Quiet.*

Place the animal in a shallow, flat-bottomed dish and leave it for a few hours or a day, in order to give it time to get accustomed to the place and come to rest on the bottom. Then, taking the utmost care not to jar the dish or breathe upon the surface of the water, look at the *Clepsine* through a low magnifying lens and see what happens when the surface of the water is touched with the point of a needle held vertically above the animal's back. If the experiment is properly carried out, it will be seen that the respiratory undulations (if such movements happen to be going on) suddenly cease and that the animal *slightly* expands its body and hugs the glass. Wait a few moments until the animal, recovering its normal composure, again resumes its respiratory movements. Then let the needle descend through the water until the point rests on the bottom of the dish at a little distance from the edge of the body. Again the movements will cease and the animal will hug the glass with its body somewhat expanded. Now push the needle slowly along towards the leech, and notice, as the needle comes almost in contact with the thin margin of the body, that the part nearest the needle begins to retreat slowly before it. This behavior shows a surprising keenness of tactile sensibility, the least touch of the water with a needle-point being felt at once. This delicate sensitiveness is manifested in such a quiet way that it would be generally overlooked, and an observer unfamiliar with the habits of *Clepsine*, and not realizing the necessity of extreme quiet in his own movements, would almost certainly draw false conclusions. If the dish were moved or the water carelessly disturbed in any way, the *Clepsine* would assume its motionless attitude and appear to be wholly indifferent to the disturbance. If its back were rubbed with a brush or the handle of a dissecting needle, in order to test its sensitiveness to touch, the appearance would probably still be that of insensibility and indifference to the treatment. Closer examination, however, would show that the flesh of the animal was more rigid than usual, and that the surface was covered with numerous small, stiff, conical elevations, the dermal papillæ or warts, which are so low and blunt in the normal state of rest as to be scarcely visible. It would be seen that the animal, although motionless, was in a state of active resistance to attack. Every muscle would be strained; the whole skin would be tense and rough with the stiff, pointed papillæ; and at the same time the body would be found excessively slippery and difficult to lay hold of, owing to the mucous secretion poured forth from the dermal glands. To guard still further against dislodgment, the body would be flattened out as much as possible and tightly applied to the glass. The activity of the resistance offered by this passive-looking creature would be very forcibly realized if the observer attempted to circumvent it by slipping a thin blade or spatula beneath it with a view to forcing its hold. If overcome in one part it would stick by another, and skillful manipulation would be necessary to get both ends free at the same time. With one end detached, the other will often hold against a pull strong enough to snap the body in two.

b. *Rolling into a Ball.*

Clepsine has another and entirely different method of keeping quiet. The behavior bears striking analogy to that which has been described as "feigning death" in some insects. The animal rolls itself up (head first and ventral side innermost) into a hard ball, outwardly passive, and free to roll or fall whithersoever gravity and currents in the water may direct it. The ball will bear considerable pressure and rough handling without unfolding or exhibiting any marked movements. Left in quiet for a few seconds, the animal slowly unrolls itself and creeps off. This instinct has many advantages for a slow-moving creature like *Clepsine*, as will presently be seen.

1. *Provoked by Exposure.* — The ball-like attitude is assumed under various circumstances. If a stone or board with *Clepsine* attached to its under surface be quietly turned upside down, thus bringing the leech from shade or darkness into light and exposure, it may sometimes maintain its position of rest unchanged, only hugging the stone a little more closely and not moving until all is quiet. More generally, however, it rolls itself up, and by the time the stone is turned, or before, it falls to the bottom, where it can unfold and escape without danger of discovery. If, by chance, the animal has eggs, it will not desert them to escape in this way. As soon as the eggs hatch and the young become attached to the ventral side of the parent, the latter may roll itself up with its brood inside, fall to the bottom as before, and thus escape with all its progeny.

This species, then, has two quite distinct and peculiar ways of keeping quiet and thus avoiding its enemies. If the animal has no eggs, or if it has young, it may adopt either mode of escape, while if it has eggs it has no choice but to remain quiet over them. In the species here considered, the eggs are held together in a thin, gelatinous sheet, secreted at the time of ovipositing, and of a size and form to be entirely covered by the expanded body of the parent. In some species of *Clepsine* the eggs are laid in thin membrane-like sacs, which are fastened to the under side of the parent, and in this case the rolling up into the form of a ball is the safest course of behavior and the one generally adopted.

2. *Forced by Attack.* — The same behavior will almost invariably follow when any species of *Clepsine* is closely pursued and finds itself unable to fix itself by either end, as when a spatula is repeatedly thrust under it in such a way as to break its hold and defeat its efforts to regain a footing.

3. *Induced by Gorging Blood.* — The provocations to such behavior thus far considered have all been such as might, and probably do, cause more or less alarm. It is important to note, however, that the instinct may manifest itself frequently under conditions that seem to exclude the influence of fear. I have often seen these leeches fold themselves into balls at the end of a good meal, and so roll to the edge of the shell of their host, and fall to the bottom. This mode of concluding a quiet repast, with no assignable cause for alarm, and with every reason for satisfaction and contentment, except for the desire to get out of light into darkness, under cover of a stone or some other object, will hardly pass as feigning; and cataplexy and the tropisms are equally out of question. We could not assume, for example, that *Clepsine* is positively heliotropic when hungry, and negatively heliotropic after feeding. *Shade is preferred at all times in both conditions.* If hungry, *Clepsine* leaves the shade, not because it prefers light, but because it prefers its host more than it prefers shade. If the host is not found it will again return to a shady retreat, if one is to be found, however hungry it may be. The rolling up cannot be attributed to light, as the animal takes the extended position when at rest, even if compelled to remain in the light. What, then, shall we conclude?

4. *Origin and Utility.* — Observation and inference may be stated as follows:

1. The act of rolling up into a passive ball may be performed (*a*) *under compulsion,* as when it is the last resort in self-defence; (*b*) *under a milder provocation,* as *one* of three courses of behavior, as when the resting place is turned up to light, and the choice is offered between remaining quiet in place, creeping away at leisure, or rolling into a ball and dropping to

the bottom ; (c) or, finally, *under no special external stimulus,* but rather *from internal motive,* the normal demand for rest and shady seclusion, presumably very strong in *Clepsine* after gorging itself with the blood of its turtle host.

2. This mode of taking leave of the turtle, after a full meal, is the easiest, the quickest, and the safest way available. To drop off fully extended, as *Clepsine* sometimes does, would retard descent and increase the chances of capture by fish. To creep about on the back of the host, waiting for an opportunity to grasp a plant or stone, would be decidedly hazardous, for if it came within the stretch of its host's neck, annihilation would be almost certain, while if lucky enough to keep out of reach of its host, it would still be in danger of the same fate from other turtles.

3. This behavior is instinctive, since it is performed by the young after the first meal as perfectly as by the adult.

4. Looking more closely at the nature and origin of this instinct, it will be seen to be quite a natural performance, in keeping with the most fundamental features of the animal's organization, and only a special application of a more general act that is primary and organic as much as tasting, seeing, or sleeping.

The more general act consists simply in tucking or rolling the head under, as often happens when the animal is resting. The habit may be observed in the young as soon as they are sufficiently developed to be capable of bending the tip of the head under. The same act, carried a little further, gives the half-rolled condition, in which only the anterior of the animal is folded, while the posterior portion remains unrolled and attached by the sucker. This attitude is often assumed if the leech is sick or has been injured. It is only a step further to release the sucker and fold it over the part already rolled up, thus completing the part ball to a whole ball, which can move passively more rapidly and safely than is possible by active creeping. From beginning to end we have only one act, in different stages of completion, simply different degrees of one and the same process.

5. Having the general act to start with, it is easy to see

how it might be made of use for particular purposes; in other words, how special adaptations of a useful kind might arise. If the act is a natural concomitant of the resting condition, and is associated with a feeling of ease and security, we see how sickness, injury, fear, a heavy meal, etc., might prompt it, and in higher or lower degree, according to the nature and intensity of the inciting cause. Full and prompt action under exposure, pressure, injury, and in the event of a good meal, would carry decisive advantages, so that individuals reacting in the more favorable degrees would stand the best chance of escape and survival. Natural selection would steadily improve upon the results, and the special adaptation, in different stages of development in different species, as we find it to-day in different *Clepsines*, would lie in the direct line of progress. This view does not of course presuppose intelligence as a guiding factor, and therefore lends no support to the theory of instinct as "lapsed intelligence," or "inherited habit."

6. An instinct of the kind here considered does not depend for its development upon effort and the transmission of functionally acquired improvements in organization, but upon *the natural selection of the best qualified germs*, for that is what the survival of the fittest individuals always means. Many species of *Clepsine* require but one full meal a year, and as they seldom live more than two or three years, the number of meals is very limited. There is little room, then, for repeating the experiment often enough to affect the organization. Indeed, such a supposition would here appear to be absurd in the last degree. On the other hand, the selection of the fittest germs, provided for in the survival of the best-adapted individuals, would inevitably advance the species along the line leading to the special instinct.

If the view here taken be correct, the instinct of rolling into a ball is not a matter of deliberation at all, but merely the action of an organization more or less nicely adjusted to special conditions and stimuli. Intelligence does not precede and direct, but the indifferent organic foundation with its general activities is primary; the special behavior or instinct is built up by slowly modifying the organic basis.

c. *Sensitiveness to Light.*

The question as to how much intelligence, if any, *Clepsine* may have, I do not here undertake to settle or discuss. That the animal is endowed with keen sensibilities is evident from the behavior before described. The following simple experiment affords a striking demonstration : Pass the hand over a dish in which a number of *Clepsines* are resting quietly on the bottom, at a distance of a few inches above the animals, taking care not to make the least jar or other disturbance. If the animals are quite hungry, the slight shadow of the hand, imperceptible though it be to our eyes, will be instantly recognized by them, and a lively scene will follow, every leech rising up, supported on its posterior sucker, and swinging at full length back and forth, from side to side, round and round, as if intensely eager to reach something. Put a turtle in the dish and see what a scramble there will be for a bloody feast. The shadow of the hand was to these creatures like the shadow of a turtle swimming or floating over them in their natural haunts, and hence their quick and characteristic response. A piece of board floating over them would have the same effect. Although so sensitive to a small difference in light, the *Clepsine* eyes can give no pictures, and hence there is no power of visual discrimination between objects. They probably recognize their right host by the aid of organs of taste; at any rate they are often able to distinguish their host from closely allied species.

INSTINCT OF ROLLING INTO A BALL AMONG INSECTS.

The following examples of the instinct of rolling into a ball among insects are from Kirby and Spence.[1]

" I possess a diminutive rove-beetle (*Aleochara complicans* K. Ms.), to which my attention was attracted as a very minute, shining, round, black pebble. This successful imitation was produced by folding its head under its breast, and turning up its abdomen over its elytra, so that the most piercing and discriminating eye would never have discovered it to be an insect. I

[1] Entomology, pp. 411, 412.

have observed that a carrion beetle (*Silpha thoracica*) when alarmed has recourse to a similar manœuvre. Its orange-colored thorax, the rest of the body being black, renders it particularly conspicuous. To obviate this inconvenience, it turns it head and tail inwards till they are parallel with the trunk and abdomen, and gives its thorax a vertical direction, when it resembles a rough stone. The species of another genus of beetles (*Agathidium*) will also bend both head and thorax under the elytra, and so assume the appearance of shining, globular pebbles.

" Related to the defensive attitude of the two last-mentioned insects, and precisely the same with that of the Armadillo (*Dasypus*) amongst quadrupeds, is that of one of the species of wood-louse (*Armadillo vulgaris*). The insect, when alarmed, rolls itself up into a little ball. In this attitude its legs and the underside of the body, which are soft, are entirely covered and defended by the hard crust that forms the upper surface of the animal. These balls are perfectly spherical, black, and shining, and belted with narrow white bands, so as to resemble beautiful beads; and could they be preserved in this form and strung, would make very ornamental necklaces and bracelets. At least so thought Swammerdam's maid, who, finding a number of these insects thus rolled up in her master's garden, mistaking them for beads, employed herself in stringing them on a thread; when, to her great surprise, the poor animals beginning to move and struggle for their liberty, crying out and running away in the utmost alarm, she threw down her prize. The golden wasp tribe also (*Chrysididæ*), all of which I suspect to be parasitic insects, roll themselves up, as I have often observed, into a little ball when alarmed, and can thus secure themselves — the upper surface of the body being remarkably hard, and impenetrable to their weapons — from the stings of those *Hymenoptera* whose nests they enter with the view of depositing their eggs in their offspring. Latreille noticed this attitude in *Parnopes carnea*, which, he tells us, *Bembex rostrata* pursues, though it attacks no other similar insect, with great fury; and, seizing it with its feet, attempts to dispatch it with its sting, from which it thus secures itself. M. Lepelletier de Saint-

Fargeau, to whom entomology is indebted for so many new facts relative to the manners of hymenopterous insects, has given us a striking account of a contest between the art of one of these parasites (*Hedychrum regium*) and the courage of one of the mason-bees in endeavoring to defend its nest from its attack.   The mason-bee had partly finished one of her cells, and flown away to collect a store of pollen and honey.   During her absence the female parasitic *Hedychrum*, after having examined this cell by entering it head foremost, came out again, and walking backwards, had begun to introduce the posterior part of her body into it, preparatory to depositing an egg, when the mason-bee arriving laden with her pollen-paste threw herself upon her enemy, which, availing herself of the means of defence above adverted to, rolled herself up into a compact ball, with nothing but the wings exposed, and equally invulnerable to the stings or the mandibles of her assailant.   In one point, however, our little defender of her domicile saw that her insidious foe was accessible ; and, accordingly, with her mandibles cut off her four wings, and let her fall to the ground, and then entering her cell with a sort of inquietude, deposited her store of food, and flew to the fields for a fresh supply ; but scarcely was she gone before the *Hedychrum*, unrolling herself, and, faithful to her instinct and her object, though deprived of her wings, crept up the wall directly to the cell from whence she had been precipitated, and quietly placed her egg in it *against the side* below the level of the pollen-paste, so as to prevent the mason-bee from seeing it on her return."

Behavior of Necturus.

a.   *Refusal of Food from Fear.*

Our large fresh-water salamander, popularly called mud-puppy, water-dog, hellbender, etc., is another animal that may be profitably studied with reference to its modes of quiet. The first adults which I kept in captivity in a large aquarium refused to eat pieces of raw beef or small fish, whether dead or alive.   For months they went on, seemingly entirely indifferent to any proffered food, not paying the least attention, so far as I noticed, to tempting morsels dropped quietly in front of them or held in suspension before them. Living earth-worms and insect larvæ were presented to them, all of which were known to be palatable to the creature in its natural habitat; but nothing availed to draw attention or elicit any evidence of hunger. Quiet and wholly indifferent in outward behavior, yet the animals were actually starving and wasting away. Were the creatures *feigning* quiet and indifference? Or was the behavior merely the expression of timidity, the animal not having the courage to perform the acts necessary to secure the food which it must have craved? I confess that I did not for a long time understand the cause of this refusal of food.

Further acquaintance with the adults, supplemented by an experience of two seasons in rearing the young, opened my eyes to the extreme timidity of these animals, which is so deep-seated and persistent that one can form only a poor idea of it without considerable actual contact with it. The outward behavior is very quiet and mild and gives little indication of fear. The animal will often submit to gentle handling without making any violent effort to escape. In short, the behavior is misleading, and one stands no chance of understanding it until he learns to keep quiet himself while observing, and discovers how to get into confidential relations with the crea-ture. This can be done with the adults, but to better advan-tage with the young.

b. *Behavior of Young in Taking Food.*

The eggs may be readily hatched in a shallow dish and young thus obtained which have never learned anything from the parents. I had about fifty young hatched in this way towards the end of July. When first hatched they were loaded with food-yolk sufficient to meet their needs for about two months. By the end of September I began to get intima-tions of a desire for food. The method of feeding was as follows: The dish containing the young was kept on a table, where, without being moved, food could be offered in perfect quiet. I used the tiniest bits of raw beef and offered only one piece at a time, which I held in small forceps or on the point of a needle a little in front of the animal to be tested. If the meat is held closely enough to touch the head, the animal is frightened and may retreat with such haste as to alarm all its companions. If the bait is held a little to one side, an inch or so away, and very quietly for a minute or more, a slight turning of the head in that direction may be noticed, in case the animal is ready to eat and feels confidence enough to try to reach it. The turning of the head is done very cautiously and almost as slowly as the minute-hand of a clock moves, so that one may become aware of it, not by seeing the movement, but by noticing the inclination of the head to the axis of the body. If there be a decided turn of the head of this kind, the case is hopeful, as it shows an interest which may be encouraged to action by bringing the bait a little nearer, but very slowly and without any jerky movement. Halting about half an inch away, wait for further movement on the part of the animal, if you are fortunate enough not to have frightened it away. If the animal's courage holds out — in most cases it does not in the first trials — it will soon begin to move, but with a slowness that tries the observer's patience. The head at length comes up to a point a quarter of an inch away, more or less, and after making sure of the position of the bait, which seems to be done less by the aid of the eyes than by the sense of touch, the animal tries to seize it by a quick side movement of the head and a snap of the jaws. *The first attempt to take the bait corresponds in all essential points with the behavior of the adult when trying to capture a fish, a worm, or an insect larva, although the aim may not be quite so sure.*

If one is successful in getting one or more to feed, the more timid ones may be brought forward in the same way by patiently alluring them from day to day, until they are tempted to an effort. Once made, the effort becomes easier at the next trial, and in the course of a month or six weeks the bolder ones will respond fairly promptly. A few of my specimens became very familiar with me, and would come towards me when I approached the dish, looking up at me wistfully, as if knowing well the meaning of my visits.

c. *Influence of Innate Timidity.*

In the behavior above described, we see an instinctive mode of capturing prey held in check, and probably directed to some extent, by innate timidity. Fear seems to be the main factor in control at the start, holding the animal in a trance-like quiet, undecided as to what to do, waiting for confidence to attack, or for a stronger motive to flee. As fear subsides a little, the preparatory movement of attack begins, but the sly behavior is due to fear rather than the slyness of stratagem. The slow and cautious method of approach is certainly not all finesse, for the deportment bears still the stamp of hesitating timidity, and this part of the act may become much freer as the animals become tamer and more fearless. The final part of the act, that of snapping the bait, was always performed in the same characteristic way. The piece of meat seemed always to be regarded as a *living* prey, which was to be seized quickly, held firmly for a moment or two, and then swallowed. Unfortunately I did not experiment to see what could be done in modifying this part of the act.

Instinctive fear is evidently a very important element in the conduct of the lower as well as the higher animals. In *Necturus* we see how it may be just as effective as intelligence in securing a sly mode of attack. So strong is its influence that I doubt whether there is any finesse in the movement. The adaptation of acts to purposeful ends must not be accepted too quickly as proof of intelligence in the doer. Such acts are common enough in plants, and there we are under the necessity of finding some other explanation.

d. *Organization Shapes Behavior.*

Necturus appears to understand well the act of capturing its prey, and the nice adaptation of each act to the end in view naturally enough suggests forethought and refined experience.

But we see the performance executed by the young, which have never had any experience of that kind, nor any opportunity to copy from others. We cannot therefore suppose that they perform these acts understandingly. The young *Necturus*, hatched in a dish where it has never met any living thing except its companions, has nothing to guide its first effort to capture food except its organization and its simple experience in walking and swimming, which acts are again like those of the adult, not because directed by intelligence or example, but because they are performed with the same organs under similar conditions. The young has the same sensory and motor apparatus as the adult, but it has never before known the feeling of hunger, it has never experienced pain from contact with an enemy, it has never learned that a prey may escape if not approached slyly. Its movements in approaching and snatching a piece of meat, as if it were a living object, are, then, those characteristic of the species, not because they are measured and adapted to a definite end by intelligent experience, but because they are organically determined ; in other words, depend essentially upon a specific organization.

The timidity of young hatched in a dish is the same as that of specimens hatched in the lake, and therefore it cannot be charged to individual experience or to parental influence. It, too, inheres in the special brand of organization, and has nothing to do with memory of pain sensations.

e. *Origin and Meaning of Behavior.*

We have taken a very important step in our study when we have ascertained that behavior, which at first sight appeared to owe its purposive character to intelligence, cannot possibly be so explained, but must depend largely, at least, upon the mechanism of organization. The origin and meaning of the behavior antedate all individual acquisitions and form part of the problem of the origin and history of the organization itself. It is the first and indispensable step, without which it would be impossible to reach sound views, either as regards the particular behavior or the difficult question of the relation of instinct to habit and intelligence. If the problem is not simplified, its nature is better defined and its perspective is relieved of many a myth that might otherwise obscure our vision. We see at once that the behavior does not stand for a simple and primary adaptation of a preëxisting mechanism to a special need. As the necessity for food did not arise for the first time in *Necturus*, the organization adapted to securing it must be traced back to foundations evolved long in advance of the species. The retrospect stretches back to the origin of the vertebrate phylum, and, indeed, to the very beginning of genealogical lines in protozoan forms. The point of special emphasis here is that instincts are evolved, not improvised, and that their genealogy may be as complex and far-reaching as the history of their organic bases.

f.   *Sensibility — Sources of Error.*

Another important factor in animal behavior, namely, sensibility, is very generally underestimated and often sadly misunderstood. We are apt to gauge sensibility according to the intensity of the overt response to stimulus, forgetting that the animal has the power to inhibit such manifestations or to moderate them in a way to mislead the observer. In the struggle for existence a high premium has been placed on this power, with the result that it is well-nigh a universal attribute. The best proof of its high value to the possessor is our own readiness to accept the disguise it affords as an evidence of lack of sensibility. We are so prone to think that the exercise of such power depends upon considerable intelligence that we are incredulous of its existence in forms that give only doubtful signs of intelligence. The power is possessed to a very marked degree by *Clepsine*, and it is only when we become aware of this fact and take all necessary precautions that we can get any reliable tests of the animal's keenness of sensibility. *Necturus* is even more difficult to manage, for not until after we have won its confidence by slow degrees can we expect free responses.

Besides the great danger of being deceived by the response, or the lack of response, to stimulus, there are two other insidious sources of error to be guarded against. We habitually assume that intelligence and sensibility rise and fall together. This idea may lead to false conclusions in two directions — to overestimating sensibility at the upper end of the scale and underestimating it at the lower. That high sensibility does not imply high intelligence is clear in the case of *Clepsine* and equally so in almost any other case that might be selected among the lowest segmented animals. That high intelligence does not necessarily imply correspondingly high sensibility is shown by the well-known fact that many animals greatly surpass man in their sense powers. It can be shown, I believe, that the difference in sensibility between higher and lower animals is very much less than is generally supposed.

The second source of error is the common assumption that the grade of sensibility rises with the structural complexity of the sense organs. This view is likewise untenable. It is true that the sense organs as a rule become more complex in structure as we go up the scale, but this advance in structure is mainly confined to accessory and non-sensory parts, which are either of a protective nature or else concerned in some subsidiary function, such as muscular adjustments and regulation of the stimuli. Such improvements in the non-sensory parts may be carried to a high state of perfection and greatly raise the general efficiency of the organ (*e.g.*, the vertebrate eye), without adding much, if anything, to the sensitiveness of the individual sense cells. The sense cells may be multiplied in number and placed in a position of safety and advantage for receiving stimuli, and the stimuli may be strengthened, directed, and otherwise regulated so as to secure the best results ; but all that may obviously not affect the functional power of the cells themselves. We do not know the range of variability in this power, but we do know that the sense cells often vary relatively little in structure, sometimes retaining in the higher forms the same typical features that characterize them in the lower forms. There is no known difference of structure that would warrant the assumption that the dermal sensillæ in annelids are less sensitive than those in aquatic vertebrates. We have, then, no reliable test of sensibility either in the structure of the sense organ, in the rank of the organism, or in its intelligence. We have to depend upon the response to stimuli, and, remembering that this may be deceptive, observe and experiment under conditions that insure *free behavior*.

No one who has never come into close communion with the lower animals can begin to appreciate the delicacy and efficiency of their sensory apparatus. We take up the earthworm and, as we see no eyes, we conclude that it cannot see. A little experiment shows that it is extremely sensitive to light, and further study of its structure reveals unpigmented eyes lying beneath the skin, and the whole surface thickly set with minute delicate tactile sensillæ. Even *Amphioxus*, so long reputed to have no visual organs, turns out to have such organs from end to end, imbedded in its spinal cord. I have before called attention to the highly sensitive organization of *Clepsine* and its allies. In the very lowest organisms, plant and animal alike, where special visual organs do not exist, the living protoplasm has, as has been demonstrated in many ways, a keen sensibility to light, so that one might look upon the whole organism as fulfilling the light-perceiving function.

g.   *Orientation through the Dermal Sensillæ.*

Necturus, as before remarked, has a very keenly sensitive organization. The skin is richly provided with sense organs, which terminate at the surface in very short, fine hairs, invisible to the naked eye. These organs, which are of the same nature and function as the dermal sensillæ in *Clepsine* and in so many other aquatic animals, are sensitive to slight vibrations in the water that are far beyond the reach of any of our sense organs, and they are the main reliance, both in avoiding enemies and seeking prey.

It is interesting to see how little the eyes are depended upon in finding a piece of meat. A bit dropped in front of a young *Necturus* receives no attention after it reaches the bottom.

An object must be in *motion* in order to excite attention, and it is not generally the moving form that is directly perceived, but the movements of the water, travelling from the object to the sensory hairs, are felt, and in such a way as to give the direction of the disturbing centre with most surprising accuracy. If a bit of beef is taken up adhering to the point of a needle, and held in the water, the vibrations imparted to the needle by the most steady hand will be sufficient to give the animal the direction. If the meat falls to the bottom, and the needle is held in place, the animal approaches the needle and tries to capture it, without paying the slightest attention to the meat lying directly below. If, after the meat has fallen, the needle is withdrawn and touched to the surface of the water behind or at one side of *Necturus,* it turns instantly in the direction of the needle, not because it sees, but because it *feels* wave motions coming from that direction. Long experience with *Necturus* and with many of its nearer allies enables me to speak very positively on this point. When it is remembered that in the higher animals the direction of sound waves is given by the auditory sense organs, which are primarily surface sensillæ homologous with those in the skin of *Necturus,* it may not seem so strange that the animal directs its movements in the way described. *Necturus* can see, but it can feel (perhaps we should say hear) so much more efficiently that its small eyes seem almost superfluous.[1]

[1] Professor Eigenmann has kindly written the following note on the use of the tactile organs in the blind fishes :

Chologaster papilliiferus, a relative of the blind fishes living in springs, detects its prey by its tactile organs, not by its eyes. A crustacean may be crawling in plain view without exciting any interest unless it comes in close proximity to the head of the fish, when it is located with precision and secured. The action is in very strong contrast to that of a sunfish, which depends on its eyes to locate its prey. A *Gammarus* seen swimming rapidly through the water and approaching a *Chologaster* from behind and below was captured by an instantaneous movement of the *Chologaster,* when it came in contact with its head. The motion brought the head of the *Chologaster* in contact with the stem of a leaf, and instantly it tried to capture this also. Since the aquarium was well lighted, the leaf in plain sight, it must have been seen and avoided if the sense of sight and not that of touch were depended upon.

In *Amblyopsis,* the largest of the blind fishes of the American caves, the batteries of tactile organs form ridges projecting beyond the general surface of the

h. *Origin and Nature of the Behavior in Taking Food.*

1. *Some Intelligence Implied.* — Let us now return to the
question of the origin and nature of the behavior of *Necturus*
in capturing its food; not, however, with the expectation of
reaching a complete solution, but rather in the hope of coming
nearer to the problem and to the guiding principles in dealing
with it.   It is obvious, first of all, that automatism will not
suffice to account for the whole behavior.   That there is
organic coördination of movements no one will dispute.   But
these movements must be steered in the direction of the
object, and this orientation does not seem to be a purely
automatic arrangement.   The dermal sensillæ ("lateral-line"
organs) give the impressions which enable the animal to
steer its course; but action and sense impression are evidently
not linked in a way to be independent of inhibitory influences.   I assume that the creature is conscious, and that it
has a certain intelligent appreciation of the sense impressions
received.   This is not saying that the young *Necturus* is a born
philosopher; I assume nothing more than that it has already
learned by experience how to *direct* its movements.   That does
not imply much, but certainly some, intelligence.   I cannot
otherwise understand why the same stimulus should not always
evoke the same response under the same conditions.   But we
see that there is hesitation about starting, and this hesitation
may be prolonged to any length, showing conclusively that
sensation and response are not so connected as to exclude

---

skin.   Its prey, since it lives in the dark and its eyes are mere vestiges, is located
entirely by its tactile organs.   This is done with as great accuracy as could be
done with the best of eyes in the light, but only when the prey is in close proximity to the head.   Coarser vibrations in the water are not perceived or are
ignored, and apparently stationary objects are not perceived when the fish
approaches them.   If a rod is held in the hand, the fish always perceives it when
within about half an inch of it, and backs water with its pectorals.   If the head
of a fish is approached with a rod, the direction from whence it comes is always
perceived and the correct motion made to avoid it.   This reaction is much more
intense in the more active young than in the adult.   One young, about 10 mm.
long, determined with as great precision the direction from which a needle was
coming as any fish with perfect eyes could possibly have done.   It reacted
properly to avoid the needle, and this without getting excited about it.

inhibitory influences. There is unmistakably a power of inhibition strong enough to counteract the strongest motive to act — the hunger of a starving animal in the presence of food.

2. *Orientation Learned by Experience.* — In assuming that the young *Necturus*, at the time of its first attempt to capture a piece of beef, has already learned to orient itself with reference to external objects, I have not gone beyond the possibilities. The animal has been out of the egg envelopes for about two months. It has been confined in a glass dish about ten inches in diameter, holding water about one inch in depth. Its life has been about as exclusively vegetative as if it had been all the while within the egg membrane, the only difference being that it has had room to straighten itself and to move about to some extent among its fellows. It has been heavily laden with food-yolk and has maintained a quiet attitude except when disturbed by change of water. Simple as the life has been, the animal has had some experience in swimming and walking, and opportunity to use to some extent its organs of orientation. The bait offered to it is something totally new in its experience, but we cannot, of course, claim that its behavior towards the bait is wholly uninfluenced by its previous experience.

3. *Deferred Instinct.* — We have to do with what Mr. Lloyd Morgan has termed a "deferred" instinct, *i.e.*, an action performed for the first time, but not until some time after birth. Mr. Morgan's remarks on the first dive of a young moor hen [1] bring out very clearly the possible influence of experience in the case of such deferred instincts.

Mr. Morgan says:

In the case of such an instinctive procedure of the deferred type as that presented by the diving of a young moor hen, though, on the first occasion of its performance, the congenital automatism predominates, *yet it is difficult to believe and is in itself improbable that the individual experience of the young bird does not, even on the first occasion, exercise some influence on the way in which the dive is performed.* If we desire to reach a true interpretation of the facts, we must realize the fact that an activity may be of mixed origin. And if we distinguish

[1] Habit and Instinct, pp. 136, 137.

— as we have endeavored clearly to distinguish — between instinct as congenital and habit as acquired, we must not lose sight of the fact that there is much interaction between instinct and habit, so that the first exhibition of a deferred instinct may well be carried out in close and inextricable association with the habits which, at the period of life in question, have already been acquired.

Although Mr. Morgan's young moor hen had undoubtedly learned far more by experience before its first dive than my young *Necturus* could have learned before its first effort to capture food, we are nevertheless well admonished to keep in mind the fact that the activity here considered may not be pure instinct. Allowing for the small though important part played by intelligence, there remains a purposive sequence of coördinated acts, which are always performed in essentially the same manner by young and old, and by the young without instruction or example or previous experience of like motive and stimulus. In so far, then, as intelligence cannot possibly be a regulating factor we must refer the activity to organization.

4. *Pause before the Bait.* — In order to exclude as much as possible the influence of experience it will be well to confine attention to the least variable part of the behavior. The concluding phase of the performance is so typical and characteristic, and so far removed from anything previously experienced, that it may be regarded as a very near approach to pure instinct. I have in mind *the pause before the bait and then the quick side-movement of the head as the jaws are opened to seize.*

If this series of acts represents an organic sequence, and if the behavior as a whole takes the form determined by the organization, as seems to me beyond reasonable doubt, we have an instinct the history of which may be coextensive with the evolution of the animal. We stand at the end of an interminable vista. The specific peculiarities of organization in *Necturus* form but an infinitesimal element of the problem. Scarcely a feature of the instinct belongs exclusively to *Necturus.* It is at least widely diffused among vertebrates, especially among fishes. The differences in the manner of execution among different forms, so far as I have observed, are of quite a superficial nature. The instinct evidently has its root in the general instinct of preying, which is doubtless coeval with animal organization. The cannibalism of our protozoan ancestors was the starting-point, and their carnal propensities were not acquired by the aid of intelligence, but given in the fundamental properties of protoplasm. The stronger ate the weaker, and made themselves stronger and more prolific by so doing. The promise of the whole animal world was contained in the act. The constitutional disposition to feed, with variable foods available, would give occasion for different appetites and various modes of getting outside of palatable victims. In primitive organisms multiplying by simple fission, structural modifications acquired during the lifetime of the individual would be carried right on from generation to generation, and hence the structural foundations for a whole animal world such as we now see could be laid in a relatively short period as compared with the time necessary to advance organization in forms limited to reproduction by germs. In fact, these fundamentals could all be established within the realm of the unicellular protozoa. Nucleus and cell-body, inner and outer layers, nerve-muscle elements, sensory and locomotor organs, mouth and stomach, respiratory and excretory mechanisms, reproductive elements, anticipating embryological development from germs — all these essentials of higher organization are presented in the protozoan.

The organic bases furnished in the protozoan world might be passed directly on to the first metazoa, or they might be reacquired in essentially the same manner as before, and in a not much longer period, as reproduction by fission would still be a condition favoring rapid organo-genesis.

To try to fill up the gaps between the protozoan and *Necturus* would lead us too far into the field of speculation, and would not contribute much to a grasp of the problem. We have to content ourselves with general facts and principles and probabilities drawn therefrom. It is enough for present purposes to know that the roots of the instinct organization we are considering run clear back to the beginnings of organo-genesis, and that they are natural products of the properties of living protoplasm. We start with known properties and get to known rudiments of organization without invoking the aid of intelligence, or finding any way in which it could be supposed to have been a guiding factor in development.

The organic basis of the preying instinct may have grown and multiplied in different phyla a long time before receiving much aid from intelligence. The rapidity and freedom of modification would be very much limited when fission ceased and reproduction by germs became the sole mode of generation. Very early in the vertebrate phylum, possibly at its dawn, the chief characters of the instinct, as we now find it, were probably fixed in structural elements differing from those in *Necturus* only in superficial details. The strikingly fish-like character of the behavior certainly suggests as much.

5. *Meaning and General Occurrence.* — If now we look more closely at the purposive character of the behavior, it will become clearer that the instinct is shared, not only by animals below *Necturus*, but also by some far above it. The pause before the final act of seizing is a well-marked feature, which means *locating the prey and fixing the aim.* The same action with the same meaning runs all through the different branches of the vertebrate phylum. It is, as I have already said, especially characteristic of the fishes and amphibia, and it is not rare among the higher branches, the reptiles, birds, and mammals. It may be seen to good advantage in the turtles, and even the common fowl halts on coming up to the insect it is pursuing in order to make sure its aim. I believe the same instinct underlies the act of *pointing* in the dog. The origin of this behavior in pointers cannot be referred to training, as was clearly seen by Darwin.[1]

It may be doubted [says Darwin] whether any one would have thought of training a dog to point had not some one dog naturally shown a tendency in this line, and this is known occasionally to happen, as I once saw in a pure terrier; the act of pointing is probably, as many have thought, *only the exaggerated pause of an animal preparing to spring on its prey.* When the first tendency to point was once displayed, methodical selection and the inherited effects of compulsory training in each successive generation would soon complete the work.

[1] Origin of Species, p. 207.

The "tendency" manifested in some one dog was regarded by Darwin as an accidental variation, the cause being unknown. May not many of the variations appearing in domestic animals, which we call "accidental," be manifestations of instinct roots of more or less remote origin?

6. *Part Played by Fear.* — We may now glance once more at the behavior as a whole, for the purpose of pointing out the part played by instinctive timidity. Gentle movements in the water, kept up with steadiness, such as are imparted by a needle in feeding as before described, may induce an attack, while less gentle or unsteady movements may lead the animal to remain quiet or to take flight. The same stimulus, according to amplitude and evenness, may then be followed either by advance, by quiet, or by retreat. In retreat, fear is manifest; in quiet it is concealed; in advance it is less concealed. There can be no doubt that fear predominates in flight and in quiet, while it certainly tempers the advance, giving the appearance of slyness deliberately acted in order to take the prey by surprise. This sly manner of advance, whatever it be due to, has a double advantage, for it is concealment against a possible foe and prevents alarming a harmless prey. If I could suppose that fear did not strongly influence the advance, I should certainly incline to think that the animal really appreciated the great advantages in quietly surprising its prey; but for reasons before given I believe the animal is quite blind to any such bearing of its action. The advantages of this manner of action, however, are just the same as if it were deliberately assumed, and the *Necturus* conducting itself in this way would certainly fare better than one reacting in a contrary way. The instincts of *Necturus* in this case coöperate to secure its welfare, while if the creature depended upon its intelligence it is difficult to see how it could escape immediate extinction.

GENERAL CONSIDERATIONS.

a. *Instinct Precedes Intelligence.*

The view here taken places the primary roots of instinct in the constitutional activities of protoplasm [1] and regards instinct in every stage of its evolution as action depending essentially upon organization. It places instinct before intelligence in order of development, and is thus in accord with the broad facts of the present distribution and relations of instinct and intelligence, instinct becoming more general as we descend the scale, while intelligence emerges to view more and more as we ascend to the higher orders of animal life. It relieves us of the great inconsistencies involved in the theory of instinct as " lapsed intelligence." Instincts are universal among animals, and that cannot be said of intelligence. It ill accords with any theory of evolution, or with known facts, to make instinct depend upon intelligence for its origin; for if that were so, we should expect to find the lowest animals free from instinct and possessed of pure intelligence. In the higher forms we should expect to see intelligence lapsing more and more into pure instinct. As a matter of fact, we see nothing of the kind. The lowest forms act by instinct so exclusively that we fail to get decided evidence of intelligence. In higher forms not a single case of intelligence lapsing into instinct is known. In forms that give indubitable evidence of intelligence we do not see conscious reflection crystallizing into instinct, but we do find instinct coming more and more under the sway of intelli-

---

[1] Professor Loeb* refers instinct back to " (1) polar differences in the chemical constitution in the egg substance, and (2) the presence of such substances in the egg as determine heliotropic, chemotropic, stereotropic, and similar phenomena of irritability." According to this view, the power to respond to stimuli lies in unorganized chemical substances, and the same powers exist in the adult as in the egg, because the same chemical substances are present. Organization serves at all stages merely as a mechanical means of giving definite directions to responses.

The view I have taken regards instinctive action as *organic* action, whatever be the stage of manifestation. The egg differs from the adult in having an organization of a very simple primary order, and correspondingly simple powers of response. Instinct and organization are, to me, two aspects of one and the same thing, hence both have ontogenetic and phylogenetic development.

* " Egg Structure and the Heredity of Instincts," *The Monist*, vol. vii, July, 1897.

gence. In the human race instinctive actions characterize the life of the savage, while they fall more and more into the background in the more intellectual races.

Every hypothesis that would derive instinctive action from teleological reflection is open to the same objections. In many cases it would be necessary to postulate an amount of prevision on the part of the animals in which the instincts arose that would simply be psychologically impossible. Conscious prevision without a possible basis in the experience of the individual, or any means of learning from others, is simply a self-contradiction. The frequently cited instance of the emperor moth puts this point in strong light. The caterpillar of this moth so constructs the upper part of its cocoon that it will resist strong pressure from without and yield to slight pressure from within. Easy egress for the imago and security against attacks from outside enemies are thus provided for. As the spinning of the cocoon happens but once in a lifetime, the caterpillar could not anticipate such needs from its own experience, nor could it learn from its parents, which were dead long before it hatched. The possibility of imitation is also excluded, as the species is not a social one.

b. *Theories of Instinct.*

1. *Pure Instinct the Point of Departure.* — The first criterion of instinct is, that it can be performed by the animal without learning by experience, instruction, or imitation. The first performance is therefore the crucial one. It is of the utmost importance in all discussion of the origin of instinct to make sure of this point, and keep clear of all ambiguous activities such as have been designated " instinct habits " (Lloyd Morgan), "acquired instincts " (Wundt), " secondary instincts " (Romanes), etc. We must not allow the question as to the relation of instinct to habit and intelligence to be obscured by confusing terminology. There may be "mixtures " and all sorts of "interactions " between habit and instinct, and these may have a far-reaching theoretical import, but they lack definiteness, and are therefore dangerous foundations for theories. A

theory of instinct must obviously make pure instinct its first concern, and keep the general course of evolution always in view.

It is not my purpose to engage in a critical examination of theories, but to indicate briefly which of the two rival theories now most in favor accords best with facts and general principles as I understand them. These two theories are the *habit theory* of Lamarck and the *selection theory* of Darwin, Wallace, and Weismann.

2. *Embryology and the Lamarckian Theory.* — The habit theory is a part of the more general theory of the transmission of acquired characters. This doctrine has never been reconciled with the teachings of embryology, the science which deals directly with the phenomena of heredity, and which is, therefore, the touchstone of every theory of inheritance. It is a fundamental tenet in embryology that all organisms reproducing exclusively by germs owe their inherited characters to the germs from which they arose, and that germs carry the primordials of adult structure, not by virtue of any mysterious transference of parental features, but by virtue of the constitution they bring with them when they arise by division of preëxisting germs. That is, I believe, a fair statement of the embryological doctrine of inheritance, which must be the final test of our theories.

The selection theory propounded by Darwin and Wallace, and further developed by Weismann, starts from the embryological law of germ continuity (Weismann), or, otherwise expressed, germ lineage, and interprets the phenomena of variation, heredity, and development, in harmony with this law and the principle of selection. This theory is incompatible with the idea that instinct is inherited habit. We could not, for example, say with Professor Wundt[1]:

"We have supposed that father can transmit to son the physiological dispositions that he has acquired by practice during his own life, and that in the course of generations these inherited dispositions are strengthened and definitized by summation."

"The occurrence [p. 405] of connate instincts renders a subsidiary hypothesis necessary. We must suppose that the physical

[1] Lectures on Human and Animal Psychology, p. 408.

changes which the nervous elements undergo can be transmitted from father to son. . . . The assumption of the inheritance of acquired dispositions or tendencies is inevitable if there is to be any continuity of evolution at all."

3. *Darwin's Refutation of Lamarck's Theory.* — Although Darwin dwelt at some length on the points of resemblance between habits and instincts, and although he thought it possible that habits could sometimes be inherited, it should be remembered that he was the first to show conclusively that "the most wonderful instincts with which we are acquainted, namely, those of the bee hive and of many ants, could not possibly have been acquired by habit " (*Origin of Species*, p. 202).   Indeed, it was he who first found in the case of neuter insects a refutation of Lamarck's doctrine of inherited habit, and at the same time a demonstration of the high efficiency of the principle of natural selection.   Darwin concludes his chapter on instinct with these memorable words :

" The case of neuter insects, also, is very interesting, as it proves that with animals, as with plants, any amount of modification may be effected by the accumulation of numerous slight, spontaneous variations, which are in any way profitable, *without exercise or habit having been brought into play.   For peculiar habits confined to the workers or sterile females, however long they might be followed, could not possibly affect the males and fertile female, which alone leave descendants.   I am surprised that no one has hitherto advanced this demonstrative case of neuter insects against the well-known doctrines of inherited habit, as advanced by Lamarck.*"

What could more forcibly illustrate the importance of crucial cases than just this work of Darwin's on the instincts of neuter insects?   Here a conclusive test is reached, and no theory of the origin of instinct can stand that disregards it.   If habit cannot possibly have had anything to do with the origin of such typical instincts, then we should at least be very cautious in appealing to it in any case.   We certainly do not want two theories to account for the same phenomenon, if one will suffice. If the theory of inherited habit is certainly false in a single case, it must be deemed false in every case, until at least it has been shown that some cases cannot be explained without it.   Is there any case where it can be clearly shown that an undoubted instinct arose from inherited habit, or any case in which it can be made clear that the theory adopted for neuter insects cannot possibly hold? Both questions, it seems to me, must be answered in the negative.

4. *Weak Points in the Habit Theory.* — The habit theory has many adherents still, and Darwin himself often found it a convenient hypothesis. But neither Darwin nor anybody else has given us a crucial test that would stand beside that furnished in neuter insects. The failure to find such a test is certainly not due to any lack of zeal or effort on the part of the advocates of the theory. The tests claimed are numerous enough, but they always fall short of the requirement. The weak points in the theory are:

1. It starts on a disputed, if not refuted assumption; namely, that habits wholly new to the individual and the species, having no hereditary basis predisposing to them, may, as the result of exercise frequently repeated, and continued in successive generations, eventually become hereditary instincts.

2. It appeals to the less typical rather than to the more typical cases — to cases in which the critical points are undetermined or doubtful, or open to a different interpretation.

3. Its definition of habit and instinct verges towards a *petitio principii.* Two or more classes of instincts are set up so as to facilitate a nearer approach to habit, *e.g.*, acquired and connate (Wundt); primary and secondary (Romanes). Habit is used indiscriminately for an action originating in some congenital variation and an action forced upon the individual by special circumstances. A fundamental distinction, on which the validity of the theory must be tested, is thus ignored.

c. *Two Demonstrations of the Habit Theory Claimed by Romanes.*

The evidence adduced to show that habit may pass into instinct cannot here be examined in detail. Romanes brings forward two cases — the instincts of *tumbling* and *pouting* in pigeons — which he declares are alone sufficient to demonstrate the theory. We may, therefore, take these as fair samples of the arguments generally appealed to.

After quoting Darwin's remarks on this subject, Romanes adds :

"This case of the tumblers and pouters is singularly interesting and very apposite to the proposition before us ; for not only are the actions utterly useless to the animals themselves, but they have now become so ingrained into their psychology as to have become severally distinctive of different breeds, and so not distinguishable from true instincts. This extension of an hereditary and useless habit into a distinction of race or type is most important in the present connection. *If these cases stood alone, they would be enough to show that useless habits may become hereditary,* and this to an extent which renders them indistinguishable from true instincts." [1]

Granting that we have here true instincts, — and I do not doubt that, — what proof have we that they originated in habits ? Did there preëxist in the ancestors of these breeds organized instinct bases, which, through the fancier's art of selective breeding, were gradually strengthened until they attained the development which now characterizes the tumblers and pouters ? Or was there no such basis to start with, but only a new mode of behavior, accidentally acquired by some one or more individuals, and then perpetuated by transmission to their offspring, and further developed by artificial selection ? The original action in either species is called a "habit," and this so-called habit must have been inherited ; ergo, habit can become instinct. Obviously, argument of that kind can have weight only with those who overlook the test-point, namely, the real nature and origin of the initial action.

If the instinct had its inception in a true habit, *i.e.,* in an action reduced to habit by repetition in the individual, and not determined in any already existing hereditary activity, is it at all credible that it could have been transmitted from parent to progeny ? Does not our general experience contradict such an assumption in the most positive manner ? But may not the habit have originated a great many times, and by repetition in successive generations, gradually have become "stereotyped

[1] Mental Evolution in Animals, p. 189.

into a permanent instinct"? To suppose that such *utterly useless* action originated a great many times without compelling conditions or any organic predisposition is not at all admissible.

Darwin saw at once from the nature of the actions that they could not have been taught, but "*must have appeared naturally*, though probably afterwards vastly improved by the continued selection of those birds which showed *the strongest propensity*." Darwin, then, postulates as the foundation of each instinct a "propensity" — something given in the constitution. That view of the matter is in entire accord with the theory adopted in the case of neuter insects and quite incompatible with the habit theory.

1. *The Instinct of Pouting.* — I believe the case is much stronger than Darwin suspected, and that it shows, not the genesis of instinct from habit, but from a prëexisting congenital basis. Such a basis of the pouting instinct exists in every dovecot pigeon, and is already an organized instinct, differing from the instinct displayed in the typical pouter only in degree. I could show that the same instinct is widely spread, if not universal, among pigeons. It will suffice here to call attention to the instinct as exhibited in the common pigeon. Observe a male pigeon while cooing to his mate or his neighbors. Notice that he inflates his throat and crop, and that this feature is an invariable feature in the act, often continued for some moments after the cooing ceases. Compare the pouter and notice how he increases the inflation whenever he begins cooing. The pouter's behavior is nothing but the universal instinct enormously exaggerated, as any attentive observer may readily see under favorable circumstances.

2. *The Instinct of Tumbling.* — The origin of the tumbling instinct cannot be fixed by the same direct mode of identification; but I believe that here also it is possible to point to a more general action, instinctively performed by the dovecot pigeon, as the probable source of origin. I have noticed a great many times that common pigeons, when on the point of being overtaken and seized by a hawk, suddenly flirt themselves directly downward in a manner suggestive of tumbling, and thus elude the hawk's swoop. The hawk is carried on by

its momentum, and often gives up the chase on the first failure. In one case I saw the chase renewed three times, and eluded with success each time. The pigeon was a white dovecot pigeon with a trace of fantail blood. I saw this same pigeon repeatedly pursued by a swift hawk during one winter, and invariably escaping in the same way. I have seen the same performance in other dovecot pigeons under similar circumstances.

But this is not all. It is well known that dovecot pigeons delight in quite extended daily flights, circling about their home. I once raised two pairs of these birds by hand, in a place several miles from any other pigeons. Soon after they were able to fly about they began these flights, usually in the morning. I frequently saw one or more of the flock, while in the middle of a high flight, and sweeping along swiftly, suddenly plunge downwards, often zigzagging with a quick, helter-skelter flirting of the wings. The behavior often looked like play, and probably it was that in most cases. I incline to think, however, that it was sometimes prompted by some degree of alarm. In such flights the birds would frequently get separated, and one thus falling behind would hasten its flight to the utmost speed in order to overtake its companions. Under such circumstances the stray bird coming from the rear might be mistaken for the moment for a hawk in pursuit, and one or more of the birds about to be overtaken be thus induced to resort to this method of throwing themselves out of reach of danger.

The same act is often performed at the very start, as the pigeon leaves its stand. The movement is so quick and crazy in its aimlessness that the bird often seems to be in danger of dashing against the ground, but it always clears every object.

As this act is performed by young and old alike, and by young that have never learned it by example, it must be regarded as instinctive, and I venture to suggest that it probably represents the foundation of the more highly developed tumbling instinct.

The behavior of the Abyssinian pigeon, which, when "fired at, plunges downwards so as to almost touch the sportsman, and then mounts to an immoderate height," may well be due to the same instinct. The noise of the gun, even if the bird were not hit, would surprise and alarm it, and the impulse to save itself from danger would naturally take the form determined by the instinct, if the instinct existed. This seems to me more probable than Darwin's suggestion of a mere trick or play.

d. *The Habit Theory Losing Ground.*

The two instincts of pouting and tumbling, claimed as demonstrations of the habit theory, thus turn out to be explicable only on the selection theory. It is significant that this theory is fast losing ground even among the psychologists. A. Forel's conversion illustrates the trend of opinion. " I, too," he says, " used to believe that instincts were hereditary habits, but I am now convinced that this is an error, and have adopted Weismann's view. It is really impossible to suppose that acquired habits, like piano playing and bicycle riding, for instance (these are certainly acquired), could hand over their mechanism to the germ-plasm of the offspring." [1]

In his latest work, *Habit and Instinct*, Lloyd Morgan has also abandoned the theory. On the same side stand James, Baldwin, Ziehn, Flügel, and others. The following, from Karl Groos, pp. 60, 61, will show how the difficulties with the theory are multiplying.

" As regards instinct," says Groos, "there is, further, the *a priori* argument that it is inconceivable how acquired connections among the brain cells could so affect the inner structure of the reproductive substance as to produce inherited brain tracts in later generations. And, finally, there is this consideration mentioned by Ziegler as a suggestion of Meynert's : ' It is well known that in the higher vertebrates acquired associations are located in the cortex of the hemispheres. As an acquired act becomes habitual, it may be assumed that the corresponding combination of nervous elements will become more dense and strong, and the tract proportionally more fixed. This being the case, it follows that the tracts of acquired and habitual association, as well as those of acquired movement, pass

---

[1] Gehirn und Seele, 1894, p. 21. Taken from The Play of Animals, by Groos, p. 56. (Translated by Elizabeth L. Baldwin.)

through the cerebrum. Instincts and reflexes, however, have their seat for the most part elsewhere. The tracts of very few of them are found in the cortex of the hemispheres. It is chiefly in the lower parts of the brain and spinal cord that the associations and coördinations corresponding to instincts and reflexes have their seat. When the comparative anatomist investigates the relative size of the hemispheres in vertebrates (especially in amphibians, reptiles, birds, and mammals), a very evident increase in size is observed which apparently goes hand in hand with the gradual gain in intelligence. In the course of long phylogenetic development, during which the hemispheres have gradually attained their greatest dimensions, they have constantly been the organ of reason and the seat of acquired association. If, then, habit could become instinct through heredity, it is probable that the cerebrum would, in much greater degree than is the fact, be the seat of instinct.' "

The stronghold of the Lamarckian view is Paleontology. It is here that the doctrine of acquired characters, or ctetology as Professor Hyatt calls it, has been nearly as unyielding as the fossils to which it adheres. But a new light seems to be penetrating even here under the name of "organic selection." This idea, first formulated by Professor Baldwin, but almost simultaneously and independently reached by Lloyd Morgan and Professor Osborn, is, that adaptive modifications are not transmissible, but that they have, nevertheless, acted as *the fostering nurses of congenital variations*, since organisms surviving through them would carry forward to the next generation such congenital variations as happened to be coincident with them. It may be, perhaps, a fine question to determine whether so-called "adaptative modifications," which really have selection value, are not themselves the coincidents of congenital bases. Be that as it may, the conversion of so eminent a paleontologist as Professor Osborn to the selection theory is all the more significant on account of the prominent part he has taken in defending the Lamarckian idea.

e. *Hyatt on Acquired Characters.*

Professor Hyatt was the first to demonstrate a wonderfully complete parallelism between the ontogenetic and the phylogenetic series, and he has presented the paleontological argument in terms that seem, to many at least, to be beyond controversy. With all respect to Professor Hyatt's monumental work, I must say that I find nothing in the evidence that compels one to take his view of acquired characters.

"We have been unable" [says Professor Hyatt] "to find any characters which were not inheritable in some series. The behavior of all characteristics which have been introduced into any series of species shows them to be subject to the law of acceleration, in whatever way they have originated, whether primarily as adaptive characters, according to our hypothesis, or by natural selection and through the combination of the sexual variations, as supposed by Weismann."[1]

This is a very sweeping statement, at least in implication. I can hardly believe that the author would have us understand that acquired characters are just as readily and invariably transmitted as congenital characters; and yet, if that is not the argument, there is no argument there. Nothing is more certain than that, in living forms accessible to direct experimental test, acquired characters are not invariably, if at all, transmissible. Demonstrations have been sought for, but so far without avail. Unless the *Arietidæ* are a wholly exceptional group, we must conclude from the above statement that all the characters found were of congenital origin, and that no acquired characters were recognized. It is easier to believe that such characters were overlooked than to believe a miracle.

The *law of acceleration* established by Professor Hyatt is complemental to *the biogenetic law* formulated by Fr. Müller and Haeckel, and both laws rest on the theory of germ continuity, as formulated by Weismann. Logically, neither of these laws implies the transmission of acquired characters. That is an assumption which has never been reconciled with the fundamental law of the genetic continuity of germs. The

[1] "Genesis of the Arietidæ," p. 43, *Mem. Mus. Comp. Zoöl.*, Cambridge, 1889.

pangenesis theory of Darwin was an attempt in this direction, but that theory has no scientific basis and it stands as a theoretical failure, rejected because it could not possibly be reconciled with what we know about the genesis of germs. That is the inevitable fate of every view which fails to adjust itself to the primary law of germ continuity.

Sense impressions and physical impressions or modifications stand on the same footing. Repetition may become habit and produce marked effects on the nervous mechanism or other organs; but the individual structure so affected is not continued from generation to generation, so that the effects are cancelled with each term of life, and there is no conceivable way by which they could be stamped upon the germs and so carried on cumulatively. If they reappear in the offspring, it cannot be because they were inherited, but because they are reproduced in the same way as they were acquired in the parent.

f. *Preformation the Essence of the Doctrine.*

This doctrine of the transmission of acquired characters is a species of preformation that eclipses the old creation hypothesis, for the miracle of stamping the germ with the form it is to present in the adult has to be repeated at each generation.

It may be objected that " stamping " is not the method by which parental characters are given to the germ. They are commonly said to be inherited. But it is too late to juggle with the term "heredity." That term either means something or nothing. If it means that characters acquired by the parent can be transmitted to the offspring, then the transmitted characters, which *ex hypothesi* are not originally determined in the germ, must in some way be determined for it by the parent. What better term than "stamp" or " impress" can be suggested? Whatever the *modus operandi*, the determining influence or impress must be imparted, at least in the great majority of cases, before development begins. Is it conceivable that perfectly definite form features can be in any way reflected back upon the ovum? Can we think of the germ as vibrating sympathetically with each acquired peculiarity of the parent organism ? What vibration could there be between germ and passive structures, such as shell configurations ? Could an indentation, groove, ridge, or protuberance forced upon a shell by environmental action be at the same time wrought into the germs in such a definite way as to reappear in the offspring without the aid of the same environmental causes ? Or could the repetition of the same environmental action on a long line of parents gradually modify the germs in the same direction ? In whatever way we turn the question, we are confronted with the same miracle of preformation. *The character arises in the parent organism by epigenesis, but it is thrown back on the germ, nobody knows or can conceive how, in such a way that it becomes a preformation capable of unfolding without the aid of its epigenetic causes.*

On the other hand, the hypothesis that all hereditary characters in organisms exclusively gamogenetic originate in spontaneous or induced (by *direct* action of environment) germ variations, appeals only to known facts and principles, and provides for the same amount of preformation as before without any miraculous transfer of characters from one organism to another. We know that germ variations are transmissible; we do not know that individually acquired modifications can be transferred to germs ; we know the principle of selection to be rational and verifiable ; we know of no substitute for it.

The Genetic Standpoint in the Study of Instinct.

a. *The Genealogical History Neglected.*

The problem of psychogenesis requires a more definite genetic standpoint than that of general evolution. It is not enough to recognize that instincts have had a natural origin; for the fact of their connected genealogical history is of paramount importance. From the standpoint of evolution as held by Romanes and others, instincts are too often viewed as disconnected phenomena of independent origin. The special and more superficial characteristics have been emphasized to the exclusion of the more fundamental phylogenetic characters.

Biologists and psychologists alike have very generally clung tenaciously to the idea that instincts, in part at least, have been derived from habits and intelligence; and the main effort has been to discover how an instinct could become gradually stamped into organization by long-continued uniform reactions to environmental influences. The central question has been : How can intelligence and natural selection, or natural selection alone, initiate action and convert it successively into habit, automatism, and congenital instinct ? In other words, the genealogical history of the structural basis being completely ignored, how can the instinct be mechanically rubbed into the ready-made organism? Involution instead of evolution; mechanization instead of organization; improvisation rather than organic growth; specific *versus* phyletic origin.

This inversion, or rather perversion, of the genealogical order leads to a very short-focussed vision. The pouting instinct is supposed to have arisen *de novo*, as an anomalous behavior, and with it a new race of pigeons. The tumbling instinct was a sort of *lusus naturæ*, with which came the fancier's opportunity for another race. The pointing instinct was another accident that had no meaning except as an individual idiosyncrasy. The incubation instinct was supposed to have arisen *after* the birds had arrived and laid their eggs, which would have been left to rot had not some birds just blundered into "cuddling" over them and thus rescued the line from sudden extinction. How long this blunder-miracle had to be repeated before it happened all the time does not matter. Purely imaginary things can happen on demand.

b. *The Incubation Instinct.*

1. *Meaning to be Sought in Phyletic Roots.* — It seems quite natural to think of incubation merely as a means of providing the heat needed for the development of the egg, and to assume that the need was felt before the means was found to meet it. Birds and eggs are thus presupposed, and as the birds could not have foreseen the need, they could not have hit upon the means except by accident. Then, what an

infinite amount of chancing must have followed before the first
"cuddling" became a habit, and the habit a perfect instinct!
We are driven to such preposterous extremities as the result
of taking a purely casual feature to start with. Incubation
supplies the needed heat, but that is an incidental utility that
has nothing to do with the nature and origin of the instinct.
It enables us to see how natural selection has added some
minor adjustments, but explains nothing more. For the
real meaning of the instinct we must look to its phyletic
roots.

If we go back to animals standing near the remote ancestors
of birds, to the amphibia and fishes, we find the same instinct
stripped of its later disguises. Here one or both parents simply remain over or near the eggs and keep a watchful guard
against enemies. Sometimes the movements of the parent
serve to keep the eggs supplied with fresh water, but aëration
is not the purpose for which the instinct exists.

2. *Means Rest and Incidental Protection to Offspring.* —
The instinct is a part of the reproductive cycle of activities,
and always holds the same relation in all forms that exhibit it,
whether high or low. It follows the production of eggs or
young, and means primarily, as I believe, *rest* with incidental
protection to offspring. That meaning is always manifest, no
less in worms, molluscs, crustacea, spiders, and insects, than
in fishes, amphibia, reptiles, and birds. The instinct makes no
distinction between eggs and young, and that is true all along
the line up to birds which extend the same blind interest to
one as to the other.

3. *Essential Elements of the Instinct.* — Every essential element in the instinct of incubation was present long before the
birds and eggs arrived. These elements are : (1) the disposition to remain with or over the eggs ; (2) the disposition to
resist and to drive away enemies ; and (3) periodicity. The
birds brought all these elements along in their congenital equipment, and added a few minor adaptations, such as cutting the
period of incubation to the need of normal development, and
thus avoiding indefinite waste of time in case of sterile or
abortive eggs.

(1) **Disposition to Remain over the Eggs.** — The disposition to remain over the eggs is certainly very old, and is probably bound up with the physiological necessity for rest after a series of activities tending to exhaust the whole system. If this suggestion seems far-fetched, when thinking of birds, it will seem less so as we go back to simpler conditions, as we find them among some of the lower invertebrate forms, which are relatively very inactive and predisposed to remain quiet until impelled by hunger to move. Here we find animals remaining over their eggs, and thus shielding them from harm, from sheer inability or indisposition to move. That is the case with certain molluscs (*Crepidula*), the habits and development of which have been recently studied by Professor Conklin.[1] Here full protection to offspring is afforded without any exertion on the part of the parent, in a strictly passive way that excludes even any instinctive care. In *Clepsine* there is a manifest unwillingness to leave the eggs, showing that the disposition to remain over them is instinctive. If we start with forms of similar sedentary mode of life, it is easy to see that remaining over the eggs would be the most likely thing to happen, even if no instinctive regard for them existed. The protection afforded would, however, be quite sufficient to insure the development of the instinct, natural selection favoring those individuals which kept their position unchanged long enough for the eggs to hatch.

(2) **Disposition to Resist Enemies.** — The disposition to keep intruders from the vicinity of the nest I have spoken of as an element of the instinct of incubation. At first sight it seems to be inseparably connected with the act of covering the eggs, but there are good reasons for regarding it as a distinct element of behavior. In birds this element manifests itself before the eggs are laid, and even before the nest is built; and in the lower animals the disposition to cover the eggs is not always accompanied by an aggressive attitude. This attitude is one of many forms and degrees. A mild self-defensive state, in which the animal merely strives to hold its position without trying to rout intruders, would perhaps

[1] *Journ. of Morph.*, vol. xiii, No. 1, 1897.

be the first stage of development. In some of the lower vertebrates the attitude remains defensive and is aggressive only in a very low degree, while in others pugnacity is more or less strongly manifested. Among fishes the little Stickleback is especially noted for its fiery pugnacity, which seems to develop suddenly and simultaneously with the appearance of the dark color of the male at the spawning season.

In pigeons, as in many other birds, this disposition shows itself as soon as a place for a nest is found. While showing a passionate fondness for each other, both male and female become very quarrelsome towards their neighbors. The white-winged pigeon (*Melopelia leucoptera*) of the West Indies and the southern border of the United States is one of the most interesting pigeons I have observed in this respect. At the approach of an intruder the birds show their displeasure in both tone and behavior. The tail is jerked up and down spitefully, the feathers of the back are raised as a threatening dog "bristles up," the neck is shortened, drawing the head somewhat below the level of the raised feathers, and the whole figure and action are as fierce as the bird can make them. To the fierce look, the erect feathers, the ill-tempered jerks of the tail, is added a decidedly spiteful note of warning. If these manifestations are not sufficient, the birds jump toward the offender, and if that fails to cause retreat, wings are raised and the matter settled by vigorous blows.

This pugnacious mood is periodical, recurring with each reproductive cycle, and subsiding like a fever when its course is run. The birds behave as if from intelligent motive, but every need is anticipated blindly; for the young pair, without experience, example, or tradition, behave like the parents.

It seems to me that this mood or disposition, although in some ways appearing to be independent of the disposition to cover the eggs, can best be understood as having developed in connection with the latter. It has primarily the same meaning, — protection to the eggs, — but the safety of the eggs and young depends upon the safety of the nest, and this accounts for the extension of its period to cover all three stages — building, sitting, and rearing.

(3) **Periodicity.** — The periodicity of the disposition to sit coincides in the main with that of the recuperative stage. Its length, however, at least in birds, is nicely correlated with, though not exactly coinciding with, the time required for hatching. It may exceed or fall short of the time between laying and hatching. The wild passenger pigeon (*Ectopistes*) begins to incubate a day or two in advance of laying, and the male takes his turn on the nest just as if the eggs were already there. In the common pigeon the sitting usually begins with the first egg, but the birds do not sit steadily or closely until the second egg is laid. The birds do not, in fact, really sit on the first egg, but merely stand over it, stooping just enough to touch the egg with the feathers. This peculiarity has an advantage in that the development of the first egg is delayed so that both eggs may hatch more nearly together. The bird acts just as blindly to this advantage as *Ectopistes* does to the mistake of sitting before an egg is laid. *Ectopistes* is very accurate in closing the period, for if the egg fails to hatch within twelve to twenty hours of its normal time, it is deserted, and that too if, as may sometimes happen, the egg contains a perfect young, about ready to hatch. Pigeons, like fowls, will often sit on empty nests, filling up the period prescribed in instinct, leaving the nest only as the impulse to sit runs down. It happens not infrequently that pigeons will go right on with the regular sequence of activities, even though nature fails in the most important stage. Mating is followed by nest-making, and at the appointed time the bird goes to the nest to lay, and after going through the usual preliminaries, brings forth no egg. But the impulse to sit comes on as if everything in the normal course had been fulfilled, and the bird incubates the empty nest, and exchanges with her mate as punctiliously as if she actually expected to hatch something out of nothing. This may happen in any species under the most favorable conditions. It is possible by giving an abundance of rich food to wind up the instinctive machinery more rapidly than would normally happen, so that recuperation may end in about a week's time, when incubation will stop and a new cycle begin, leading to the production of a second set of eggs in the same nest.

This has happened several times with the crested pigeon of Australia (*Ocyphaps lophotes*).

Schneider[1] says : " The impulse to sit arises, as a rule, when a bird sees a certain number of eggs in her nest." Although recognizing a *bodily disposition* as present in some cases, sitting is regarded as a *pure perception impulse.* I hold, on the contrary, that the *bodily disposition* is the universal and essential element, and that sight of the eggs has nothing to do primarily with sitting. It comes in only secondarily and as an adaptation in correlation with the inability in some species to rear more than one or two broods in a season. In such species the advantage would lie with birds beginning to incubate with a full nest.

The suggestions here offered on the origin of the incubation instinct, incomplete and doubtful as they may appear, may suffice to indicate roughly the general direction in which we are to look for light on the genesis of instincts. The incubation instinct, as we now find it perfected in birds, is a nicely timed and adjusted part of a periodical sequence of acts. If we try to explain it without reference to its physiological connections in the individual, and independently of its developmental phases in animals below birds, we miss the more interesting relations, and build on a purely conjectural chance act that calls for a further and incredible concatenation of the right acts at the right time and place, and is not even then completed until its perpetuation is secured by a miracle of transmission.

A Few General Statements.

1. Instinct and structure are to be studied from the common standpoint of phyletic descent, and that not the less because we may seldom, if ever, be able to trace the whole development of an instinct. Instincts are evolved rather than involved (stereotyped by repetition and transmission), and the key to their genetic history is to be sought in their more general rather than in their later and incidental uses.

[1] Der Thierische Wille, pp. 282, 283. As cited in Professor James's Psychology, p. 388.

2. The primary roots of instincts reach back to the constitutional properties of protoplasm, and their evolution runs, in general, parallel with organogeny. As the genesis of organs takes its departure from the elementary structure of protoplasm, so does the genesis of instincts proceed from the fundamental functions of protoplasm. Primordial organs and instincts are alike few in number and generally persistent. As an instinct may sometimes run through a whole group of organisms with little or no modification, so may an organ sometimes be carried on through one or more phyla without undergoing much change. The dermal sensillæ of annelids and aquatic vertebrates are an example.

3. Remembering that structural bases are relatively few and permanent as compared with external morphological characters, we can readily understand why, for example, five hundred different species of wild pigeons should all have a few common undifferentiated instincts, such as, drinking without raising the head, the cock's time of incubating from about 10 A.M. to about 4 P.M., etc.

4. Although instincts, like corporeal structures, may be said to have a phylogeny, their manifestation depends upon differentiated organs. We could not, therefore, expect to see phyletic stages repeated in direct ontogenetic development, as are the more fundamental morphological features, according to the biogenetic law. The main reliance in getting at the phyletic history must be comparative study.

5. Instinct precedes intelligence both in ontogeny and phylogeny, and it has furnished all the structural foundations employed by intelligence. In social development also instinct predominates in the earlier, intelligence in the later stages.

6. Since instinct supplied at least the earlier rudiments of brain and nerve, since instinct and mind work with the same mechanisms and in the same channels, and since instinctive action is *gradually* superseded by intelligent action, we are compelled to regard instinct as the actual germ of mind.

7. The automatism, into which habit and intelligence may lapse, seems explicable, in a general way, as due more to the preorganization of instinct than to mechanical repetition. The

habit that becomes automatic, from this point of view, is not an action on the way to becoming an instinct, but action preceded and rendered possible by instinct. Habits appear as the uses of instinct organization which have been learned by experience.

8. The suggestion that intelligence emerges from blind instinct, although nothing new, will appear to some as a complete *reductio ad absurdum*. But evolution points unmistakably to instinct as nascent mind, and we discover no other source of psychogenetic continuity. As far back as we can go in the history of organisms, in the simplest forms of living protoplasm, we find the sensory element along with the other fundamental properties, and this element is the central factor in the evolution of instinct, and it remains the central factor in all higher psychic development. It would be strange if, with this factor remaining one and the same throughout, organizing itself in sense organs of the keenest powers and in the most complex nerve mechanisms known in the animal world — it would be strange, I say, if, with such continuity on the side of structure, there should be discontinuity in the psychic activities. Such discontinuity would be nothing less than the negation of evolution.

9. We are apt to contrast the extremes of instinct and intelligence—to emphasize the blindness and inflexibility of the one and the consciousness and freedom of the other. It is like contrasting the extremes of light and dark and forgetting all the transitional degrees of twilight. In so doing we make the hiatus so wide that derivation of one extreme from the other seems about as hopeless as the evolution of something from nothing. That is the last pit of self-confounding philosophy.

Instinct is blind; so is the highest human wisdom blind. The distinction is one of degree. There is no absolute blindness on the one side, and no absolute wisdom on the other. Instinct is a dim sphere of light, but its dimness and outer boundary are certainly variable; intelligence is only the same dimness improved in various degrees.

When we say instinct is blind, we really mean nothing more than that *it is blind to certain utilities* which we can see. But we ourselves are born blind to these utilities, and only discover them after a period of experience and education. The discovery may seem to be instantaneous, but really it is a matter of growth and development, the earlier stages of which consciousness does not reveal.

Blindness to the utilities of action no more implies unconsciousness in animals than in man. It is the worst form of anthropomorphism to claim that animal automatism is devoid of consciousness, for the claim rests on nothing but the assumption that there are no degrees of consciousness below the human. If human organization is of animal origin, then the presumption would be in favor of the same origin for consciousness and intelligence. Automatism could not exclude every degree of consciousness without excluding every form of organic adaptation.

10. The clock-like regularity and inflexibility of instinct, like the once popular notion of the "fixity" of species, have been greatly exaggerated. They imply nothing more than a low degree of variability under normal conditions. Discrimination and choice cannot be wholly excluded in every degree, even in the most rigid uniformity of instinctive action. Close study and experiment with the most machine-like instincts always reveal some degree of adaptability to new conditions. This was made clear by Darwin's studies on instincts, and it has been demonstrated over and over again by later investigators, and by none more thoroughly than by the Peckhams in the case of spiders and wasps.[1] Intelligence implies varying degrees of freedom of choice, but never complete emancipation from automatism. The fundamental identity of instincts and intelligence is shown in *their dependence upon the same structural mechanisms and in their responsive adaptability.*

INSTINCT AND INTELLIGENCE.

In order to see how instinctive action may graduate into intelligent action it is well to study closely animals in which the instincts have attained a high degree of complexity, and

[1] Wisconsin Geological and Natural History Survey, Bulletin No. 2, 1898.

in which there can be no doubt about the automatic character of the activities. These conditions are perfectly fulfilled in the pigeons, a group in which we have the further advantage that wild and domestic species can be studied comparatively.

It is quite certain that pigeons are totally blind to the meanings which we discover in incubation. They follow the impulse to sit without a thought of consequences; and no matter how many times the act has been performed, no idea of young pigeons ever enters into the act.[1] They sit because they feel like it, begin when they feel impelled to do so, and stop when the feeling is satisfied. Their time is generally correct, but they measure it as blindly as a child measures its hours of sleep. A bird that sits after failing to lay an egg, or after its eggs have been removed, is not acting from " expectation," but because she finds it agreeable to do so and disagreeable not to do so. The same holds true of the feeding instinct. The young are not fed from any desire to do them any good, but solely for the relief of the parent. The evidence on this point cannot be given here, but I believe it is conclusive.

But if all this be true, where does the graduation towards intelligence manifest itself. Certainly not in a comprehension of utilities which are discoverable only by human intelligence. Whatever the pigeon instinct-mind contains, it is safe to say that the intelligence is hardly more than a grain hidden in bushels of instinct, and one may search more than a day and not find it.

a. *Experiment with Pigeons.*

Among many tests, take the simple one of removing the eggs to one side of the nest, leaving them in full sight and within a few inches of the bird on the nest. The bird sees the uncovered eggs, but shows no interest in them; she keeps

---

[1] Professor James, Psychology, II, p. 390, thinks such an idea may arise and that it may encourage the bird to sit. " *Every instinctive act in an animal with memory*," says James, "*must cease to be ʻ blind ʼ after being once repeated.*" That must depend on the kind of memory the animal has. It is possible to have memory of a certain kind in some things, while having absolutely none of any kind in other things. That is the case in pigeons, as I feel very sure.

her position, if she is a tame bird, and after some moments begins to act as if the current of her feelings had been slightly disturbed. At the most she only acts as if a little puzzled, as if she realized dimly a change in feeling. She is accustomed to the eggs, and now misses them, or, rather, misses something, she knows not what. Although she does not know or show any care for the eggs out of the nest, she does appear to sense a difference between having and not having.

There is, then, something akin to memory and discrimination, and little as this implies it cannot mean less than some faint adumbration of intelligence. Now this inkling of intelligence, or, if you prefer, this nadir of stupidity, so remote from the zenith of intelligence, is not something independent of and foreign to instinct. It is instinct itself just moved by a ripple of change in the environment. The usual adjustment is slightly disturbed, and a little confusion in the currents of feeling arises, which manifests itself in quasi-mental perplexity. That is about as near as I can get to the contents of the pigeon mind without being able, by a sort of metempsychosis suggested by Bonnet, to live some time in the head of the bird.

In this feeble perplexity of the pigeon's instinct-mind, in this "nethermost abyss" of stupidity, there is a glimmer of light, and nature's least is always suggestive of more. The pigeon has no hope of graduating into a *homo sapiens*, but her little light may flicker a little higher, and all we need to know is, how instinct behavior can take one step toward mind behavior. This is the dark point on which I have nothing really new to offer, although I hope not to make it darker.

b. *The Step from Instinct to Intelligence.*

Some notion of what is involved in the step may be gathered by comparing wild with semi-domesticated and fully domesticated species. These grades differ from each other in respects that are highly suggestive. In the wild species the instincts are kept up to the higher degrees of rigid invariability, while in species under domestication they are reduced to various degrees of flexibility, and there is a correspondingly greater freedom of action, with, of course, greater liability to irregularities and so-called "faults." These faults of instinct, so far from indicating psychical retrogression, are, I believe, the first signs of greater plasticity in the congenital coördinations and, consequently, of greater facility in forming those new combinations implied in choice of action.

If we place the three grades of pigeons under the same conditions and test each in turn in precisely the same way, we can best see how domestication lets down the bars to choice and at the same time gives more opportunities for free action. The simplest experiment is always the best. Let us take three species at the time of incubation and repeat with each the experiment of removing the eggs to a distance of two inches outside the edge of the nest. The three grades are well represented in the wild passenger pigeon (*Ectopistes*), the little ring-neck (*Turtur risorius*), and the common dovecot pigeon (*Columba livia domestica*). The results will not, of course, always be the same, but the average will be about as follows :

1. *The Passenger Pigeon.* — The passenger pigeon leaves the nest when approached, but returns soon after you leave. On returning she looks at the nest, steps into it, and sits down as if nothing had happened. She soon finds out, not by sight, but by feeling, that something is missing. Her instinct is keenly attuned and she acts quite promptly, leaving the nest after a few minutes without heeding the egg. The conduct varies relatively little in different individuals.

2. *The Ring-neck Pigeon.* — The ring-neck is tame and sits on while you remove the eggs. After a few moments she moves a little and perhaps puts her head down, as if to feel the missing eggs with her beak. Then she may glance at the eggs and appear as if half consciously recognizing them, but make no move to replace them, and after ten to twenty minutes or more leave the nest with a contented air, as if her duty were done ; or, she may stretch her neck toward the eggs and try to roll one back into the nest. If she succeeds in recovering *one*, she is satisfied and again sinks into her

usual restful state, with no further concern for the second egg. The conduct varies considerably with different individuals.

3. *The Dovecot Pigeon.* — The dovecot pigeon behaves in a similar way, but will generally try to get *both* eggs back; and, failing in this, she resigns the nest with more hesitation than does the ring-neck.

4. *Results Considered.* — The passenger pigeon's instinct is wound up to a high point of uniformity and promptness, and her conduct is almost too blindly regular to be credited even with that stupidity which implies a grain of intelligence. The ring-neck's stupidity is satisfied with one egg. The dovecot pigeon's stupidity may claim both eggs, but it is not always up to that mark.

In these three grades the advance is from extreme blind uniformity of action, with little or no choice, to a stage of less rigid uniformity, with the least bit of perplexity and a very feeble, uncertain, dreamy sense of sameness between eggs *in* and eggs *out* of the nest, which prompts the action of rolling the eggs back into the nest. That is the instinctive way of placing the eggs when in the nest, and the neck is only a little further extended in drawing the eggs in from the outside. How very narrow is the difference between the ordinary and the extraordinary act! How little does the pendulum of normal action have to swing beyond its usual limit![1]

But this little is in a forward direction, and we are in no doubt as to the general character of the changes and the modifying influences through which it has been made possible. Under conditions of domestication the action of natural selection has been relaxed, with the result that the rigor of instinctive coördinations which bars alternative action is more or less

[1] We come to equally surprising results in many different ways. Change the position of the nest-box of the ring-neck, without otherwise disturbing bird, nest, or contents, and the birds will have great difficulty in recognizing their nest, for they know it only as something in a definite position in a fixed environment. If a pair of these birds have a nest in a cage, and the cage be moved from one room to another, or even a few feet from its original position in the same room, the nest ceases to be the same thing to them, and they walk over the eggs or young as if completely devoid of any acquaintance with or interest in them. Return the cage to its original place and the birds know the nest and return to it at once.

reduced. Not only is the door to choice thus unlocked, but more varied opportunities and provocations arise, and thus the internal mechanism and the external conditions and stimuli work both in the same direction to favor greater freedom of action.

When choice thus enters, no new factor is introduced. There is greater plasticity within and more provocation without, and hence the same bird, without the addition or loss of a single nerve-cell, becomes capable of higher action and is encouraged and even constrained by circumstances to *learn* to use its privilege of choice.

Choice, as I conceive, is not introduced as a little deity, encapsuled in the brain. Instinct has supplied the teleological mechanism, and stimulus must continue to set it in motion. But increased plasticity invites greater interaction of stimuli and gives more even chances for conflicting impulses. Choice runs on blindly at first, and ceases to be blind only in proportion as the animal learns through nature's system of compulsory education. The teleological alternatives are organically provided; one is taken and fails to give satisfaction; another is tried and gives contentment. This little freedom is the dawning grace of a new dispensation, in which education by experience comes in as an amelioration of the law of elimination. This slight amenability to natural educational influences cannot, of course, work any great miracles of transformation in a pigeon's brain; but it shows the way to the open door of a freer commerce with the eternal world, through which a brain with richer instinctive endowments might rise to higher achievement.

The conditions of amelioration under domestication do not differ in kind from those presented in nature. Domestication merely bunches nature's opportunities and thus concentrates results in forms accessible to observation. Natural conditions are certainly working in the same direction, only more slowly. The direction and the method of progress must, in the nature of things, remain essentially the same.

Nature works to the same ends as intelligence, and to the natural course of events I should look for just such results as

Lloyd Morgan [1] so clearly pictures and ascribes to intelligence. "Suppose," says Mr. Morgan, "the modifications are of various kinds and in various directions, and that, associated with the instinctive activity, a tendency to modify it *indefinitely* be inherited. Under such circumstances *intelligence would have a tendency to break up and render plastic a previously stereotyped instinct.* For the instinctive character of the activities is maintained through the constancy and uniformity of their performance. But if the normal activities were thus caused to vary in different directions in different individuals, the offspring arising from the union of these differing individuals would not inherit the instinct in the same purity. The instincts would be imperfect, and there would be an inherited tendency to vary. *And this, if continued, would tend to convert what had been a stereotyped instinct into innate capacity ; that is, a general tendency to certain activities (mental or bodily), the exact form and direction of which are not fixed, until by training, from imitation or through the guidance of individual intelligence, it became habitual. Thus it may be that it has come about that man, with his enormous store of innate capacity, has so small a number of stereotyped instincts.*"

The following from Professor James [2] is suggestive :

" Nature implants contrary impulses to act on many classes of things, and leaves it to slight alterations in the conditions of the individual case to decide which impulse shall carry the day. Thus, greediness and suspicion, curiosity and timidity, coyness and desire, bashfulness and vanity, sociability and pugnacity seem to shoot over into each other as quickly, and to remain in as unstable equilibrium, in the higher birds and mammals as in man. They are all impulses, congenital, blind at first, and productive of motor reactions of a rigorously determinate sort. Each one of them, then, is an instinct, as instinct is commonly defined. *But they contradict each other; experience, in each particular opportunity of application, usually deciding the issue. The animal that exhibits them loses the 'instinctive' demeanor and appears to lead a life of hesitation and choice, an intellec-*

---

[1] Animal Life and Intelligence, pp. 452, 453.
[2] Psychology, II, pp. 392, 393.

tual life; not, however, because he has no instincts — rather because he has so many that they block each other's path."

Looking only to the more salient points of direction and method in nature's advance towards intelligence, the general course of events may be briefly adumbrated. Organic mechanisms capable of doing teleological work through blindly determined adjustments, reproduced congenitally, and carried to various degrees of complexity and inflexibility of action, were first evolved. With the organization of instinctive propensities, liable to antagonistic stimulation, came both the possibility and the provocation to choice. In the absence of intelligent motive, choice would stand for the outcome of conflicting impulses. The power of blind choice could be transmitted, and that is what man himself begins with.

Superiority in instinct endowments and concurring advantages of environment would tend to liberate the possessors from the severities of natural selection ; and thus nature, like domestication, would furnish conditions inviting to greater freedom of action, and with the same result, namely, that the instincts would become more plastic and tractable. Plasticity of instinct is not intelligence, but it is the open door through which the great educator, experience, comes in and works every wonder of intelligence.

Spencer[1] has shown clearly that this plasticity must inevitably result from the progressive complication of the instincts. *" That progressive complication of the instincts,"* he says, *" which, as we have found, involves a progressive diminution of their purely automatic character, likewise involves a simultaneous commencement of memory and reason."*

---

[1] Psychology, I, pp. 443 and 454, 455.

8

HERBERT SPENCER JENNINGS
1868–1947

Jennings pursued his undergraduate work at the University of Michigan, then graduate study at Harvard under Edward Laurens Mark. There he received his Ph.D. in 1896 with work on the morphogenesis of a rotifer. A year in Jena and a visit to Naples completed his study. The year 1897–1898 he taught at Montana State University (then the Agricultural College of Montana in Bozeman), then the next year at Dartmouth. In the several years at Harvard, when he spent the summers working for the United States Fish Commision, Jennings may have come to the attention of some of the MBL staff. Charles Davenport, then at Harvard, also encouraged Jennings. As the MBL neurology seminar continued into the late 1890s, Jennings was invited to lecture.

Following the example of work in comparative psychology by Edward Thorndike, Jennings' experimental techniques led him to conclude that organic nature has basic reactions which occur throughout all levels of the animal world. Organisms do not behave as they do because of physicochemical tropisms, as Loeb maintained, according to Jennings. The reactions are not dependent on specific types of stimuli, but are more general. Jennings's protozoa were more primitive than Loeb's metazoa and therefore might well provide more evidence about the most basic nature of organisms. Because of that, as well as the weight of the evidence which he continued to accumulate until his book, *Behavior of the Lower Organisms*, appeared in 1906, Jennings offered a major challenge to Loeb's dominance. The elements of that challenge are already evident in his lecture to the MBL. There he acknowledged the indeterminate nature of "positive chemotaxis," for example, and set up the study which followed over the next few years into the workings of responses to chemical and other stimuli.

HERBERT SPENCER JENNINGS, WITH HIS WIFE.

THE BEHAVIOR OF UNICELLULAR ORGANISMS

HERBERT S. JENNINGS

IN recent biological writings there is manifest a growing tendency to interpret the processes taking place within the bodies of higher animals — especially the developmental processes — as a series of responses to stimuli. In the egg and the developing embryo, masses of protoplasm migrate from one position to another, cells and cell masses alter in form, changes of the most varied character are continually occurring. To explain such changes it is becoming usual to call upon chemotaxis, geotaxis, phototaxis, thigmotaxis, and other motor reactions of similar character. The prevalent ideas of these reactions, known usually under names terminating in *-taxis* or *-tropism*, have been derived to a large extent from the phenomena shown by the movements of unicellular organisms; the classic experiments of Pfeffer on the chemotaxis of bacteria and flagellates and of Strasburger on the phototaxis of swarm spores having opened a fountain from which all have felt entitled to draw. To understand the migration of a cell or mass of cells in the embryo we are referred back to experiments on unicellular organisms, wherein it is shown that the movements of the latter are controlled by chemical agents, by heat, by light, and the like. Here the vital processes are seemingly brought into the closest relation with chemical and physical ones; chemotaxis, for example, is frequently interpreted as the direct expression of chemical affinity or chemical repulsion between the substance of the protoplasmic mass and some other substance, or between two protoplasmic masses. There is thus established an immediate direct relation between the movements of organisms and movements characteristic of inorganic substances; a long step is taken toward that analysis of vital processes into simple chemical and physical ones, which is deemed by many the final goal of biological science. If these phenomena do indeed establish such a relation, they challenge the attention of every man interested in the fundamental phenomena of life; in any case, they invite complete and thorough investigation of the claims made for them. Since it is largely from the reactions of free unicellular organisms that our ideas of chemotaxis, phototaxis, and the like have been derived, it is important to study carefully the reactions of these creatures and to determine the laws which control them. We shall then be in a position to decide whether the movements of these organisms do furnish a key to the understanding of ontogenetic processes or not. It is these considerations that have impelled the investigation whose main results I shall try to present.

In studying the behavior of single-celled creatures we are forced into relation with the much debated question of the nature and importance of the activities of unicellular organisms as compared with those of higher animals and plants. Some hold that the cellular standpoint is the fruitful one for general physiology; that we must first determine the laws of action for single cells, then carry these over to the cell state, understanding the latter only as a combination of the former. Some go so far as to maintain that the reactions of unicellular organisms are of an intrinsically different character from those of higher forms, being of essentially the same nature as the reactions of inorganic bodies ; this is, for example, the position of Le Dantec.[1] Others hold that the division of organisms into cells is, physiologically at least, a secondary matter; that nothing more fundamental is to be expected from the study of a unicellular organism than from that of one composed of many cells. This question can be decided, of course, only by a thorough study of both the classes of organisms thus contrasted, with a comparison of the results, to see if the study of the simpler organisms does, as a matter of fact, clear up

[1] La matière vivante, Chapters I and II.

and simplify the phenomena exhibited by the many-celled creatures.   For one desirous only of getting at the real laws underlying the phenomena, the conflict on such points between high
authorities [1] is very confusing, and the only recourse is to a
first-hand study of the facts.

In the hope of getting light on the problems proposed and
others of similar character, I shall set forth and discuss observations and experiments made upon a number of free-swimming
unicellular organisms.   In the investigation it was found well
to begin with some single species and work out its activities,
and the laws governing the same, completely enough to reveal
their essential nature, then to make a comparative study of the
activities of other organisms in the light of the knowledge so
gained.   The same method will be advantageous for the presentation of results.

I give first, therefore, some of the results of a preliminary
study of the activities of *Paramecium caudatum*.   This is one
of the commonest of the ciliate Infusoria, living by thousands
in vegetable matter decaying in water.   It is a somewhat cigar-
shaped creature, having a broad groove passing obliquely from
one end (the anterior) to the mouth, which lies at about the
middle of the length of the body.   The side on which the
mouth and groove lie may be called the *oral* side; the opposite
one the *aboral* side.   The entire surface of the animal is covered
with cilia, by means of which Paramecium moves.

In beginning a study of the activities of such an organism,
we are at once confronted with the question of its psychic
powers.   If these unicellular organisms do, as a matter of fact,
possess so complicated and highly developed a psychic life as
Binet, in his book on the *Psychic Life of Micro-Organisms*, has
attempted to show obtains among them, then indeed there is
little prospect of gaining light on simple migrations of protoplasmic masses during development, through a study of their
behavior.   A study of the chick or the dog would perhaps be
as promising.   The activities which Paramecium shows are at

[1] See, for example, Verworn, "General Physiology," and Loeb, "Einige
Bemerkungen über den Begriff, die Geschichte und Literatur der allgemeinen
Physiologie," *Pflüger's Archiv*, Bd. lxix, p. 249.

first view of great complexity, so that they might seem to entirely justify Binet's views as to the height and variety of the psychic powers of these organisms. These activities and their explanation have been discussed somewhat fully by the writer in a paper [1] devoted entirely to the psychological aspect of the matter, so that only so much of this aspect will be taken up at present as has a necessary relation to the questions proposed.

If we place a number of Paramecia, in the culture water in which they are found, upon a glass slide, and cover with the cover glass, we soon find that the animals, which were at first scattered uniformly, have gathered into groups in one or more parts of the preparation. Usually we find that a bit of bacterial zoöglœa forms the center of such a group; as many of the Paramecia as can do so have pressed their anterior ends against the mass, the ciliary current carrying bacteria to their mouths; others press in from behind. It is well known, of course, that Paramecia make no choice in the food which the current brings to their mouths, taking in particles of all sorts indiscriminately. The possibility may suggest itself, however, that they have gathered about these masses of zoöglœa because the latter serve them as food. The choice of food would thus occur a step sooner — the Paramecia choosing their food by gathering about it, then taking whatever comes. But if we introduce into the slide a bit of filter paper or a fine raveling of cloth, we find that the Paramecia gather about it with the same apparent avidity as about the zoöglœa, pressing the anterior end against it and remaining thus, quiet, for long periods.

This and other experiments show, therefore, that this gathering about a bit of bacterial zoöglœa or other substance is not the expression of a choice of food, but is merely a manifestation of the fact that the Paramecia react to contact with solids of a certain physical texture by suspending active locomotion and remaining against the solid. A similar reaction to solids is, of course, a very common phenomenon among organisms of different sorts; it has received the name "Thigmotaxis," or "Stereotropism."

[1] "The Psychology of a Protozoan," *Amer. Journ. of Psychology*, vol. x, No. 4, 1899.

We have in thigmotaxis one of the fundamental reactions of Paramecium, not further analyzable into simpler component reactions. As it seems to consist chiefly or entirely of a cessation of a part of the usual ciliary motion, — only the cilia in the oral groove continuing to strike strongly backward, — it may be more philosophical to consider this partly resting condition as the "normal" condition, the usual forward motion being then considered a reaction to a stimulus, due to a change or removal of the solid bódy against which the animal is resting, or to some other change in the environment. There seems to be no decisive reason for considering either the condition of partial rest or of the usual forward motion as more "normal" than the alternative condition; taking either as a starting point, the other may be considered a response to a stimulus.

If the Paramecia are placed upon the slide in pure water, containing no bacterial zoöglœa, or any other solid, they do not even then remain scattered uniformly throughout the preparation. On the contrary, it is usually not long before the animals are gathered into one or more close groups in some part of the slide. Paramecia are usually found in the culture jars also aggregated into groups; this, taken together with the above experimental demonstration that Paramecia, at first uniformly scattered, will soon collect into close groups without evident external cause, might be held to indicate the existence of a "social instinct" among these creatures. Another possibility suggests itself — that there may be some invisible chemical substance in the region of these groups by which all the Paramecia are attracted; so that the fact that they come near together would be a secondary result of the fact that all are attracted by the same substance.

The main results of the extended study of the conduct of the Paramecia toward chemicals, to which this possibility led, may be given in a few words. It was found that Paramecia tend to gather together and form collections in drops of weakly acid solutions, and in solutions of some salts, while they avoid alkaline solutions and solutions of the salts of the alkali metals.

Among the substances into solutions of which they gather is carbon dioxide. If a bubble of carbon dioxide is introduced into a preparation of Paramecia, they soon collect closely about it and swim in circles around it without leaving it.

It was then proved, by introducing the Paramecia into a solution of rosol, which is decolorized by carbon dioxide, that these Infusoria excrete a distinctly appreciable amount of this substance, which diffuses into the surrounding water. Whenever, therefore, a very few Paramecia get together, an active solution of carbon dioxide is soon formed, and the region becomes at once a center of attraction for the Paramecia. A most complete correspondence was demonstrated between the diffusion of the CO_2 into the water and the distribution of the Paramecia in groups, and all the phenomena exhibited by the (apparently) spontaneous collections of Paramecia could be exactly imitated by introducing CO_2 into the slide.

Thus these collections of Paramecia give no indication of "social instinct," but are merely the expression of positive chemotaxis on the part of the animals toward a certain substance. In the same way all the seemingly complex activities of these creatures may be reduced to simple factors, so that there seems no evidence to indicate the possession by them of psychic powers of anything more than the most elementary character.

We may proceed then to a closer analysis of the apparent attractions and repulsions — chemotaxis, thermotaxis, and the like; it is from a study of these that light is to be gained on the problems first proposed. We shall first consider chemotaxis.

The fact that animals and plants are attracted by certain chemical substances and repelled by others is of course well known for a large number of organisms. As to the essential nature of this phenomenon, opinions differ. As pointed out above, some hold that chemotaxis is the direct expression of chemical affinity or repulsion between the living protoplasm and the chemical. Le Dantec (*La matière vivante*, pp. 51, 52) gives geometrical figures illustrating the action between the surface of a free cell and a chemical substance diffusing in the surrounding water, demonstrating in mathematical form that as a result of this action the cell must move either toward or away from the center of diffusion of the chemical. The motion of such a protoplasmic body would be passive in the same sense as the movements of iron filings are passive when acted upon by a magnet.   Delage and Herouard[1] actually state that the Flagellata have, in addition to their usual active movements, also a *passive* motion, due to the attraction of chemical substances.   Perhaps the majority of biologists hold less radical views than this; yet the opinion seems widespread that in chemotaxis we are dealing with a simple primary phenomenon.

Coming now to an examination of the phenomena as exhibited by Paramecium, we will first take up positive chemotaxis, or attraction toward chemical substances.   The phenomenon to be explained shows itself as follows.   If into a slide of Paramecia a drop of some attractive substance (as a weak acid) is introduced, the Infusoria soon collect in the drop, forming there a dense assemblage.   Now, what is the exact action of the attractive substance on the Paramecia to cause them to turn and enter the drop?   Observing carefully the conduct of the animals, we find, first, that *they do not* turn toward the drop. Owing to its slow diffusion, the margin of a drop thus introduced beneath the cover glass is evident, and the Paramecia, swimming in every direction throughout the preparation, may be seen in their random course to graze almost the edge of the drop, without their motion being changed in the least; they keep on straight past the drop and swim to another part of the slide.   But of course some of the Paramecia in their random swimming come directly against the edge of the drop.   These do not change their motion, but keep on undisturbed across the drop.   But when they come to the opposite margin, where they would if unchecked pass out again into the surrounding medium, a marked reaction is caused; the Paramecium jerks back and turns again into the drop.   Such an animal then swims across the drop in the new direction till it again comes to the margin, when it reacts negatively, as before.   This continues, so that the animal appears as if caught in the drop as in a trap. Other Paramecia enter the drop in the same way and are imprisoned like the first, so that in time the drop swarms with the animals.   As a result of their swift, random movements when

[1] Traité de zoologie concrète, tome i, p. 305.

first brought upon the slide, almost every individual in the preparation will in a short time have come by chance against the
edge of the drop, will have entered and remained, so that soon
all the Paramecia in the preparation are in the drop.

Thus it appears that the animals are not attracted by the
fluid in the drop; they enter it by chance, without reaction,
then are repelled by the surrounding fluid. The peculiar fact
that the animals, after entering the drop of the substance in
question, are repelled by the surrounding fluid in which they
were previously immersed will become more comprehensible
after the phenomena* of repulsion are considered.

Turning, then, to the matter of negative chemotaxis or
repulsion, we have the following phenomenon to be explained.
If into a slide of Paramecia swimming at random a drop of some
repellent chemical (as NaCl) is introduced, we find that the
drop remains entirely empty, not a single Paramecium entering
it. Now, exactly how do the Paramecia succeed in keeping out
of such a repellent solution?

Careful observation shows that when the Paramecium, swimming forward, comes in contact with the drop of repellent substance, it swims backward a short distance, then turns *toward
its own aboral* side, then swims forward again. The essential
point in this reaction method is, that the Paramecia always turn
toward their own aboral side, without regard to the position of
the stimulating drop. If a Paramecium comes obliquely in
contact with the drop so as to touch it only on one side of its
body, it nevertheless gives the reaction above described without modification, even though turning toward its own aboral
side after backing off may carry the animal directly toward the
drop, instead of away from it. In such a case the animal when
it comes again in contact with the drop simply repeats the
reaction. As it continually revolves on its long axis both when
swimming forward and when swimming backward, the aboral
side is nearly certain to lie in a new position the second time,
so that the animal turns in a new direction. If this is repeated
a sufficient number of times, the Paramecium is fairly certain,
by the laws of chance, to get started finally in a direction which
carries it away from the stimulating chemical.

It thus appears that the direction in which a Paramecium turns after stimulation by a chemical substance is not determined by the position of the stimulating agent, nor indeed by any external factor, but by an internal factor, — by structural differentiations of the animal's body.   This is demonstrated in a striking manner by immersing the Paramecia directly into a chemical solution of such a nature as to act as a stimulus. The entire surface of the animal is then bathed by the chemical, so that there is nothing in the external conditions to determine in which direction the animal shall move.   Nevertheless, under these circumstances, it swims backward, turns toward the aboral side, and swims forward, usually repeating the operation indefinitely.   Very striking is also the experiment of causing the chemical to act first upon the posterior end of the animals.   This may be done as follows : A large number of Paramecia are frequently observed with anterior ends pressed against the surface of a bit of bacterial zooglœa (thigmotactic reaction), so that the posterior ends are all pointed in the same direction.   Now, a capillary glass rod, coated with some chemical, is introduced into the water behind the Paramecia.   The chemical gradually diffuses through the water, of course first reaching the posterior ends of the Paramecia.   But these, when they respond, react exactly as in the other cases ; they swim backward some distance and turn toward the aboral side.   It often occurs that in thus swimming backward they enter the densest part of the chemical and are killed by it.

These experiments indicate that not only the direction of turning after swimming backward, but also the swimming backward itself is determined by internal factors, and is independent of the position of the source of stimulus.   This conclusion seems strictly true for chemical stimuli both in Paramecium and in other Infusoria experimented with.   As will be shown later, other experiments throw a light upon the cause of this uniform backward motion when stimulated by a chemical.

Summing up, then, we may say that when Paramecium is chemically stimulated it swims backward, turns toward its own aboral side, then swims forward.   As a rule, the anterior end, moving forward, comes first in contact with the chemical, so that swimming backward does, as a matter of fact, usually carry the animal away from the source of diffusion of the chemical, and turning toward the aboral side before swimming forward again will, as a rule, if repeated, finally carry the animal in such a direction that it does not again come against the source of stimulus. But these are, from the physiological standpoint, matters of accident; the animal conducts itself in the same way whether the source of stimulus has this usual position at the anterior end of the animal or not. The direction of motion after a chemical stimulus, then, has no relation to the position of the chemical substance. We cannot say, therefore, that the Paramecia are *repelled* by any chemical substance — just as we were compelled to conclude that they are not directly attracted by any chemical substance.

We find, then, that the effect of chemicals on Paramecia is not to attract or repel them, but simply to cause a certain set formula of movements. Such a set formula of movements, "touched off," as it were, by stimuli of various sorts, may be called a reflex. In returning now to the question of how the apparent attraction of the Paramecia toward certain substances — that is, the fact that they collect in drops of certain substances — can be brought about through such a reflex, it is necessary to recall certain general facts in regard to the nature of reflexes. First, any change in the environment that can be perceived by the organism may "touch off" such a reflex. Second, the character of the reflex has no necessary relation to the nature of this external change, so that of a given kind of change it cannot be predicted beforehand whether it will cause the reflex or not, and changes of opposite character may produce the same reflex.

The mechanism of the gathering together of the Paramecia into a drop of some weak acid is then as follows : When the Paramecium passes from the surrounding fluid into the acid solution there is, of course, at the moment of crossing the boundary of the drop a change in its environment. Whether this change will cause the characteristic reflex or not is impossible to predict, that depending upon the internal mechanism of the organism ; as a matter of fact we find that it does *not*

cause the reflex.   Now, after passing across the drop it comes
again to the boundary where, if not stopped, it would pass out
again into the surrounding fluid.   At this boundary there is, of
course, another change in the environment — a change in the
opposite sense from that experienced in passing into the drop.
Whether this second change will cause the reflex is of course
likewise impossible to predict, since it depends upon the nature
of the organism ; as a matter of fact we find that it *does* cause
the reflex.   The Paramecium is, therefore, returned into the
drop and kept there in the manner already described.   It
seems probable that the physiological condition of the Paramecium is changed by immersion in the drop of acid, so that
contact with the culture fluid now acts as a stimulus, though it
before did not.   It seems not impossible to conceive, however,
that even without such a change in physiological condition, an
environmental change from b to a might cause a reaction, when
the opposite change, from a to b, would cause none.   This has,
as is evident from the nature of a reflex, no necessary relation
to the comparative actual mechanical difficulty in passing in
one direction or the other.

The one effect of a marked chemical stimulus on Paramecium
is, then, to produce the characteristic reflex already described,
and the apparent attraction or repulsion is determined by the
fact that some chemical substances or chemical changes cause
the reaction, while others do not.

Now, experimentation with stimuli other than chemical leads
to the highly important observation that this same reflex is
produced by stimuli of the most varied nature.   Substances
which seem to act upon Paramecium only through their osmotic
pressure, such as solutions of sugar, cause the same reflex ;
tonotaxis, then (to use the name employed by Massart), acts
through the same reflex as does chemotaxis.   Mechanical stimuli, produced by jarring the preparation, cause the same reflex.
Heat and cold act in the same way, and the Paramecia avoid
hot or cold areas and collect in regions of optimum temperature
in exactly the same manner as they avoid certain chemicals and
collect in others.

We are driven, therefore, to the conclusion that chemotaxis

is not an activity differing in kind from the other reactions of these animals. Many sorts of changes in the environment produce a certain characteristic reflex in Paramecia, resulting in their collecting in regions of certain characters and leaving other regions vacant. Among the changes that act thus are chemical changes, and the characteristic groupings of the animals so caused are said to be due to chemotaxis; they are, however, produced in an essentially similar manner to the groupings produced by other agents. There is a unity underlying the motor activities of the Paramecia — a unity expressed in the fact that the different classes of stimuli produce identically the same reaction.

To be accurate, however, we must distinguish two less important forms of reaction to stimuli that are not manifested through the characteristic reflex above described. One is thigmotaxis; this is, however, not a motor reaction, but one characterized chiefly or entirely by a cessation of a part of the usual motion. Again, as previously set forth, it is possible to consider the partially resting condition characteristic of thigmotaxis as the primary condition; then the ordinary forward motion of the animal will be a motor reaction to a stimulus, since it is induced by a change in the environment. As will be shown, there is sufficient ground in certain other Infusoria to *compel* us to consider this forward motion as at times a reaction to stimulus; this, then, is a motor reaction which does not take place through the above-described characteristic reflex. It seems possible that the following represents the true state of the case; very weak stimuli acting on the resting individual cause the ordinary forward motion; stronger stimuli produce the above-described motor reflex.

In view of the means by which chemotaxis is brought about, it becomes more intelligible why the Infusoria may at times collect in regions of injurious substances and avoid at times areas of harmless substances. It is not a matter of attraction or repulsion at all. In the former case the injurious substance merely does not act as a stimulus to cause the motor reflex; in the second case, the chemical in question, though not injurious, does act as a stimulus. An extended investigation directed upon this point showed that the chief factor determining whether a substance does or does not cause the motor reflex of Paramecium is not the injuriousness of the substance, but is of a chemical nature.

We are now prepared to sum up the main results on Paramecium. In this animal we find that chemotaxis, thermotaxis, tonotaxis, reactions to mechanical shock, and the like, are not distinct kinds of activity; that in each case we have the same movements, merely induced by different agents. When Paramecium is effectively stimulated by any substances acting chemically or through osmosis, by heat or by cold or by mechanical shock, it responds with a reflex, which consists of the following activities: the animal swims backward, turns toward its own aboral side, then swims forward. The result of this method of reaction is that the Paramecia tend to leave the sphere of influence of agents causing this reflex, and to congregate in areas where this reaction is not caused. For chemical substances at least it is proved that the position of the stimulating agent has no influence on the direction of movement after a stimulus; the direction of movement throughout the reaction is determined by internal factors.

Is this reaction method one that is common among unicellular organisms, or is it peculiar to Paramecium? To answer this question I have studied the reactions of a considerable number of unicellular organisms belonging to the Flagellata and Ciliata. The essential point in the reaction of Paramecium, the factor that gives character to the entire response, is the circumstance that the animal after stimulation turns toward one side which is structurally defined, without regard to the nature and position of the source of stimulus. The point to which attention was primarily directed in studying the other organisms was, therefore, whether after stimulation the creature turned always toward one structurally defined side.

The organisms studied included, among the flagellates: *Chilomonas paramecium* and *Euglena viridis;* in the ciliates the following Holotricha: *Paramecium caudatum, Loxophyllum meleagris, Colpidium colpoda, Microthorax sulcatus, Dileptus anser, Loxodes rostrum,* and a species of *Prorodon;* the following Heterotricha : *Stentor polymorphus, Spirostomum ambiguum,* and *Bursaria truncatella ;* of Hypotricha, *Oxytricha fallax* and a number of undetermined species.   In several of these creatures, on account of the large size or other favorable circumstances, it was possible to use methods of investigation not available for Paramecium ; in particular it was possible in a number of cases to localize very precisely the action of stimuli.

In all of the organisms named, in spite of great variations in the nature and complexity of the usual movements, the reaction method was essentially similar to that of Paramecium.   In all, the direction of turning after a stimulus was toward a structurally defined side, without regard to the nature and position of the source of stimulus.   With regard to the details of the reaction, as might be expected, the greatest variety exists, but the general reaction plan was the same throughout.

This method of reaction evidently has a close relation to the usual asymmetry of the cell body exhibited by these organisms. This asymmetry of the Infusoria has also a close relation to the normal method of progression through the water, as well as to the method of reaction to a stimulus.   Most of these organisms, as they swim forward, also revolve on the long axis, and the resulting path is usually a spiral.   The form of the body has a constant relation to the axis of the spiral, the same side being at all times directed toward this axis.   The unsymmetrical structure of the body, the usual method of progression, and the method of reaction to a stimulus are thus evidently closely interrelated.   In the case of a bilaterally or radially symmetrical animal one would certainly not expect that one side would be always preferred to the other in turning away from a source of stimulus, as is the case in the Infusoria.

In the case of chemical stimuli it was found for all the organisms studied that not only the turning to one side, but the swimming backward after a stimulus, was independent of the position of the source of diffusion of the chemical.   The action of chemical stimuli was localized by bringing a capillary glass rod coated with some chemical compound near the anterior end, one side, or the posterior end, of the quiet organisms.   In every case (except in *Euglena viridis,* which does not swim backward under any circumstances) the organisms reacted to the chemical stimulus by swimming backward, turning toward the usual structurally defined side, then swimming forward. The swimming backward, of course, sometimes carried the creature away from the densest part of the solution (when the chemical was held in front); at other times, directly toward and into the densest part (when the same chemical was held behind). In the latter case the organisms were frequently killed by swimming into the dense solution. Thus, in chemical stimuli, without exception, the direction of motion after stimulation has no relation to the localization of the stimulus.

In several of the organisms it was possible to use also very precisely localized mechanical stimuli; and the results so gained tend to modify in some particulars the general conclusions that might be drawn from a study of the action of localized chemical stimuli. Localized mechanical stimuli were produced by touching under a powerful lens any desired part of the body of the organism with a glass rod drawn to the finest hair in a flame. For Paramecium itself this method of experimentation was not satisfactory, owing to the minute size of the cell body. One point of importance was brought out in Paramecium, however. The anterior tip of the body was shown to be incomparably more sensitive than any other part. On bringing the glass hair near the anterior tip, Paramecium leaps backward almost before the hair is seen to have reached it, giving the entire typical reaction already described. Any other part of the body was so insensible that it was not possible to cause a reaction of any sort by touching it with the hair. Paramecium could be pushed about and made to alter its direction of movement mechanically, of course, but there was no active response of any sort when it was touched at any point except the anterior end.

In *Spirostomum ambiguum* essentially the same results were reached with mechanical as with chemical stimuli. If any part of the body was touched, whether anterior end, posterior end, or side, the infusorian gave the typical reaction — swimming backward, turning toward the aboral side, then swimming forward. A slightly greater percentage of cases of the typical reaction was obtained by touching the anterior end, but the difference was little; it varied in Spirostoma from different cultures.

In the other organisms on which the effects of localized mechanical stimuli were tried, particularly *Loxodes rostrum, Dileptus anser, Oxytricha fallax*, and one or two other Hypotricha, the following results were obtained: (1) The side toward which the animal turns after a stimulus is entirely independent of the side which is touched. In every case the organism turns toward one structurally defined side. If that is the side which is touched, the organism turns continually toward the source of stimulus, no matter how many times the latter is repeated; if the other side is touched, the creature of course turns away from the source of stimulus. The impression is given that it is physiologically impossible for the organism to turn otherwise than toward this one side. (2) But the forward or backward movement of the animals after a stimulus is *not* thus independent of the localization of the stimulus. If the anterior end is touched, the organism darts backward, turns toward one side, then swims forward. The posterior half of the body is very insensible, so that as a rule there is no response to a mechanical stimulus occurring here. If, however, a strong stimulus is given here, as by thrusting the tip of the rod strongly against the resting animal, the latter simply *swims forward;* if already swimming forward, it merely hastens its forward motion when thus stimulated.

Thus, in the case of mechanical stimuli in these organisms the direction of motion after a stimulus depends, *to a certain extent*, so far as backward or forward motion is concerned, upon the localization of the stimulus. This introduces a greater complexity into the psychology of these creatures than the results on Paramecium alone, or on the reactions to chemical stimuli alone would lead us to judge. The organisms do in certain respects react with reference to the localization of a stimulus affecting them. The differing results gained with chemical stimuli are probably to be interpreted, in view of the facts shown by a study of mechanical stimuli, as follows: When a chemical diffuses from a point lying behind the infusorian, it of course comes first in contact, as a very weak solution, with the posterior end of the animal. Now, as already stated, this posterior end is not at all sensitive, so that no reaction is caused. The chemical continues to diffuse until it finally reaches the very sensitive anterior end, when at once the typical reaction occurs, and the animal swims backward into the strong solution. The reaction to a chemical is perhaps then always due to stimulation at the anterior end.

Psychologically considered, we seem to have here a remarkable transitional condition toward a perception of the localization of the stimulus by the organism — a reaction with reference to the localization of the stimulus so far as motion along the axis of the body is concerned, a blind reflex, without regard to the localization of the stimulus, so far as motion to one side is concerned.

We may now summarize briefly the essential facts in regard to the reactions of the unicellular organisms studied. The reactions of these organisms may be classified into three reaction forms:

(1) One is the thigmotactic reaction. Starting with the moving infusorian, we find that it reacts to contact with solid bodies of a certain physical texture by suspending part of the usual ciliary motion, so that locomotion ceases and the organism remains pressed against the solid. Whether anything more than this cessation of part of the usual ciliary motion is concerned in the thigmotactic reaction is very difficult to say.

(2) If we start with the resting individual, the simplest reaction to a stimulus is the resumption of the usual forward motion. This is the reaction that is produced when the solid substance against which the creature is resting is removed; it is also produced in some Infusoria when the posterior part of the body is stimulated mechanically.

(3) The third, and, for our purpose, most important reaction, to which most of the so-called tactic or tropic phenomena are due, may occur in either active or resting animals. It is a reflex consisting of the following activities: the animal swims backward, turns toward one structurally defined side, then swims forward. This reaction is produced by chemical stimuli acting upon any part of the body or upon the entire body at

once, by osmotic stimuli, by heat, by cold, by mechanical shock. Its general effect is to take the organism out of the sphere of operation of the agent causing the stimulus, and to prevent it from reëntering. The fact that certain areas are left vacant is because the agencies within these areas cause this reaction ; the collecting of the organisms within certain areas is due to the fact that here the reaction is not produced, while it *is* caused, by the influences active in the surrounding regions.

Thus, chemotaxis, tonotaxis, thermotaxis, and the like are unified ; they are not qualitatively different activities, but are fundamentally one activity due to different causes.   The tactic phenomena of unicellular organisms are brought throughout under the same point of view as the motor reflexes so well known in the physiology of higher animals.

We may now return to a brief consideration of the problems which formed the starting point of this investigation — the relation of the phenomena studied to the growth processes in the protoplasmic masses of higher organisms.   Do the laws of the motor reactions of unicellular organisms, chemotaxis and the like, really give us a basis for the understanding of protoplasmic migrations and other processes in growth and differentiation ?

We find that the tactic phenomena of these unicellular forms are brought about through a reflex that is in all essential points similar to the reflexes of higher animals.   The nature of this reflex is closely bound up with the physiological and structural differentiations of the body of these organisms ; it has a specially close relation to the asymmetry of the cell body in these Protozoa, and to the manner of the usual forward motion. These differentiations are absent in the masses of protoplasmic substance that are moved about in the processes taking place within the eggs and embryos of Metazoa.   It is difficult to see how the laws controlling the movements of such substance masses can have any similarity to the laws above developed for the reflexes of free unicellular organisms.   Above all, it is evident that the tactic movements of unicellular organisms are not direct expressions of simple chemical and physical laws ; chemotaxis, for example, is not a direct result of chemical affinities and repulsions between the protoplasmic substance and other chemical compounds. Like all the other tactic phenomena, it is the result of a motor reflex, which may be produced by the most varied means. These tactic movements, then, do not establish an immediate relation between the movements of organisms and the movements characteristic of inorganic substances. The organism reacts as an individual, not as a substance. To my mind the facts above brought out in regard to the movements of these creatures tend, if these facts have a general validity, to deprive such movements of their supposed value for explaining or illustrating the processes of growth; in so far as the ideas of chemotaxis and the like in growth processes have been derived from the phenomena exhibited by unicellular organisms, these ideas require a revision.

Especially do the facts above brought out reveal the fallacy of the statement so often insisted upon, that the growth processes induced by chemical or physical agencies are "the same as" or "identical with" the locomotor reactions induced by the same agencies. This has been carried so far that strenuous objection has been raised even to the use of distinguishing terms for these two sets of phenomena. We are told that to distinguish as *-taxis* the motor reactions of a free organism from *-tropism* or the growth reactions of a fixed organ or organism is all wrong ; the two are "identical." It is reasonably certain that the growth phenomena of plants are not brought about through a reflex that is identical with the motor reflex of Paramecium ; it seems exceedingly probable that the ways by which movements are brought about as responses to stimuli in the various classes of plants and animals will present great variety. It is difficult to see what is to be gained except confusion of ideas by applying the same names to two such dissimilar activities as the motor reflex of Paramecium when stimulated by a chemical, and the bending of a plant to or from a chemical in solution.

In regard to the second question touched upon in my introduction, — the nature and importance of the activities of unicellular organisms as compared with those of many-celled creatures, and their value for explaining the phenomena shown by the higher, — the general trend of the answer is, I think, evident. I should be inclined to interpret the facts presented somewhat as follows : The claim that the motor processes of unicellular organisms form a connecting link between inorganic processes and the vital phenomena of higher creatures clearly receives no justification for the organisms studied. Every influence coming in from outside passes, as it were, through a sort of central station, where it is completely transformed to appear as a reflex action, the nature of which is conditioned by the form and structure of the organism ; and the steps in the transformation are no more evident than they are in the higher forms. The reactions of these creatures are indeed simple, but not qualitatively of a different sort from those of higher organisms, so that for motor reactions of the sort studied I do not see that a knowledge of the conduct of these particular unicellular organisms really adds to our insight into the causal relations in the activities of higher animals.

On the other hand, if we dismiss any idea of getting from them knowledge of a different kind from that gained by the study of other groups, then the behavior of these Protozoa is of the greatest interest from the standpoint of comparative psychology. In these creatures we see, as nowhere else, how activities that seem so complicated and varied as to require psychological powers of a high order, are produced merely through one or two simple reflexes ; it seems not impossible that the phenomena exhibited in the conduct of these organisms may in time furnish important points of support for the general theory of the origin and development of psychic powers.

9

THOMAS HUNT MORGAN
1866–1945

Morgan's Ph.D. from Johns Hopkins in 1890 and his following year as a Bruce Fellow carrying out research provided him with a solid background in morphological studies. During his Hopkins years, he attended the Chesapeake Zoological Laboratory sessions in Jamaica and Bermuda and there received his initiation into marine work, which he continued at Woods Hole. The summers of 1888 and 1889 he spent at the United States Fish Commission, then 1890 and most years thereafter at the MBL, becoming an MBL Trustee in 1896.

Morgan's first work was in traditional descriptive morphology, detailing the life and development of sea spiders. His years at Bryn Mawr and at the MBL took him into contact with Jacques Loeb, who stimulated Morgan to explore new problems. Indeed, Morgan often received inspiration from others such as Loeb and was a genius at recognizing productive problems and methods. Thus, after his visit to the Naples Zoological Station in 1894–1895, where he met and worked with Hans Driesch in experimental embryology, Morgan began to explore more experimental problems. Eventually he turned to the effects of hypertonic sea water on development, work inspired also by Loeb's analysis of artificial stimulation of cleavage and eventually artificial parthenogenesis. These studies all directly influenced his continuing concentration on regeneration. For many years, Morgan examined regeneration of various organisms, generally searching for clues as to how an organism gains its differentiated form under normal conditions. Presumably, experimentally stimulated regeneration would illuminate what happens normally. Morgan acknowledged the preliminary or hypothetical state of knowledge by 1898 and 1899 when he lectured to the MBL and by 1901 when he published his volume *Regeneration*. Yet he believed that he had good evidence to support at least a working hypothesis to explain regeneration.

THOMAS HUNT MORGAN.

REGENERATION: OLD AND NEW INTERPRETATIONS

T. H. MORGAN

THE great interest that was awakened in the last century in the study of regeneration was the result of the experiments of Trembley, Réaumur, Bonnet, and Spallanzani. The interest aroused through the work of these four naturalists has not decreased, although from time to time other problems have come to the front and attracted the attention of investigators. More particularly in our own time has attention been directed to problems connected with the egg and its development. But it is becoming clearer, I think, that development by means of an egg and development by means of regeneration cannot be considered as separate and different phenomena, but at bottom have many factors in common. There can be little doubt that the results in one of these fields of study will throw light on the other. It will be possible for me to consider at this time only the suggestions and hypotheses advanced in connection with the problems of regeneration. At another time I shall try to compare these interpretations with those connected with the development of the egg.

Trembley's experiments on Hydra were the starting-point for the three other naturalists. Spallanzani occupied himself mainly with collecting new facts, while his friend Bonnet, who also made many new observations of great value, seems to have been more interested in the theoretical side of the problem.

Bonnet supposed that regeneration was brought about through the development of preformed germs. These germs exist in the animal solely for the purpose of replacing lost parts, and since in some animals the same part could be replaced time after time, Bonnet assumed that on each occasion a new set of germs was awakened. He pointed out that, since some animals are more subject to injuries than others, they are supplied with as many sets of germs as the times the animal is liable to be injured during its natural life.

Bonnet seems to have been especially impressed by the fact that from the same region of Lumbriculus a head or a tail may arise according to whether that region happens to lie at the anterior or posterior end of the cut surface. For instance, if the worm is cut into two pieces, a new tail will develop from the posterior end of the anterior piece, and a new head from the anterior end of the posterior piece. If, however, the cut had been made a little further in front of or behind this level, the same result would have followed; hence it is clear that at every level a head or a tail may develop. Which of these develops is determined by the position of the region, i.e., whether it lies at the exposed anterior or posterior end of a piece. Bonnet interpreted this to mean that there are throughout the worm head germs and tail germs. He saw that it is necessary to give some further explanation of why the one rather than the other kind of germ is aroused to activity, and made, therefore, a further assumption. The fluids of the body that pass forward carry nourishment for the head. When the worm is cut in two, these substances are, in the posterior piece, stopped in their forward movement by the cut surface, and accumulate at this place. They act especially on the head germ, and, nourishing it, bring about the development of the new head. Similarly, fluids to nourish the tail are assumed to flow posteriorly; hence in an anterior piece of a worm they will accumulate at its posterior cut surface, and, acting on the tail germ, there bring about the development of a new tail.

In another species of fresh-water annelid Bonnet found (1745) that when the worm was cut in two a new tail developed at the anterior end of the posterior piece, and not a head. He supposed that in this worm only tail germs are present throughout the body; hence only a tail can develop, even at the anterior end of the piece.[1]

In one passage Bonnet states that the fluids that flow towards the head are there used up, and we must infer that these head-nourishing fluids are being continually made somewhere else in the body of the worm. It may be pointed out, in passing, that this idea of Bonnet's, that the fluid passing towards the head (he seems to have had the blood in view) is a special kind of fluid laden with head-nourishing substances, is not in agreement with what we know of the function of the blood or of other fluids in the body. The tissues of the head may take out of the blood those substances in it that they use in their life processes, but the blood itself going to the head is not specialized in a particular direction, and is the same fluid that flows posteriorly in other vessels.

Bonnet advanced three ideas: preformed germs; head- and tail-nourishing stuffs; and the flow of these latter in definite directions. I shall return later to these views and consider them more fully.

The process of regeneration has been often compared to the completion of a broken crystal; just as the growth of an animal or of a plant is sometimes contrasted with the growth of a crystal in a saturated solution. Herbert Spencer, in particular, has elaborated this view. In his book on the *Principles of Biology*,[2] he says: "What must we say of the ability an organism has to re-complete itself when one of its parts has been cut off? Is it of the same order as the ability of an injured crystal to re-complete itself? In either case new matter is so deposited as to restore the original outline. And if in the case of the crystal we say that the whole aggregate exerts over its parts a force which constrains the newly integrated molecules to take a certain definite form, we seem obliged in the case of the organism to assume an analogous force." Starting here with an hypothesis that is no longer held, *viz.*, "that the whole

---

[1] Bonnet does not tell us how, in this case, the germs are awakened, since tail-stimulating fluids are assumed to flow backwards. Perhaps, only tail germs being present, he did not think it necessary to apply his subsidiary hypothesis.

[2] Chapter IV, Waste and Repair. First published in 1863. I quote from the last edition, 1898.

aggregate exerts over its parts a force," etc., Spencer follows this up with the *non sequitur*, "we seem obliged, in the case of the organism, to assume an analogous force."

After showing that this property, *i.e.*, "this tendency to aggregate into specific forms," cannot reside in the chemical compounds, *because*, if it resided in "the molecules of albumen or fibrin or gelatine or other proteid," there would be nothing to account for the unlikeness of different organisms ; and after showing further that it cannot reside in the morphological units or cells, *because* the same power is shown by unicellular organism, Spencer concludes that this proclivity is "possessed by certain intermediate units which we may call physiological," etc.

Striking as is the comparison between the growth of a broken crystal and the regeneration of an injured animal or plant, the emptiness and superficiality of the comparison are at once apparent on closer examination. The looseness of Spencer's argument is equally evident, and his rejection of the idea that the chemical substances composing protoplasm cannot account for the facts leaves him only his imaginary physiological units to bring about the results. The latter, after all, could only be formed by combinations of the chemical substances, and if so, why introduce into the argument unknown "units" having the property of bringing about regeneration ? To my mind nothing could confuse the whole subject more surely than reasoning and arguments like those advocated by Spencer.

The recent work of Rauber on the "regeneration" of crystals gives us now a basis of fact on which to rest any comparison we may make. Rauber's results show that during the growth of a broken crystal the typical form may be assumed and the broken surfaces obliterated. In some cases the growth may be more rapid over the broken surface, since this rougher surface presents a greater area for the deposition of new molecules. When a piece is broken off, the closing in of the exposed part presents no phenomena that are in any way different from the growth of the crystal everywhere else. The position of the new molecules that are added is determined by the condition at the points to which they are applied. It is misleading to speak of the "whole aggregate exerting over its parts a force," etc.

Rauber himself speaks with much reserve in his comparison between the "regeneration of the crystal" and the regeneration of animals and plants. His results show, it seems to me, very clearly that the comparison is only a general analogy; and the moment we attempt to press the comparison further it breaks down at every point.

It will, no doubt, be admitted by every one at the present time that the form of the crystal is somehow determined by the chemical composition of the substance of which the crystal is made up, and likewise that the form of an animal is also determined by the substances of which it is composed; but the broken crystal regains its original form only when surrounded by a saturated solution containing the same substance as the crystal. New substance is then added over its entire surface as well as over the part broken off. On the other hand, the process of regeneration is entirely different in an animal or plant. The new material, if any is formed, comes from within the animal or plant. Further, an animal that is slowly starving and decreasing in size will regenerate a missing part. Again, several of the lower forms regenerate by changing over the entire piece into the typical form. Can any one suppose for a moment that such a process is comparable to the re-completion of the form of a crystal?

If further evidence is asked, attention may be drawn to those cases in which an organ, different in kind from the one cut off, is regenerated. A striking case of this sort is the regeneration of an antenna in place of an eye, or the development of a tail at the anterior end of a posterior piece of an earthworm.

It would not be difficult to multiply at length these points of difference, but what I have said will suffice, I hope, to show how little we can gain by Spencer's comparison.

Pflüger [1] has given a brief outline of his conception of the process of regeneration. He says, since there is always

---

[1] Pflüger, E. Ueber den Einfluss der Schwerkraft auf die Theilung der Zellen und auf die Entwicklung des Embryos, pp. 64-67. 1883.

replaced only as much as is lost, it is clear that the new limb (of a salamander) does not arise from a preëxisting germ of a limb. The wounded surface of the limb draws food material to itself, and these new food molecules are organized into a new limb. The arranging force is a molecular force that does not work at a distance from the living substance of the stump of the limb, but acts only in so far that the force draws the food molecules into the sphere of influence of the molecules in the stump itself, bringing them to a definite place, and in this way laying down a new living layer over the cut end of the limb. The way in which this new layer is organized depends on the law of organization of the part, *i.e.*, on the chemical condition of the surface layer on which the new layer is deposited. In a word, the condition of the new layer is the necessary mathematical consequence of the condition of the older regenerating layer.

On the surface of the wound a knob of indifferent tissue arises, but long before the first layer of the new material has become fully formed (differentiated?) it has in turn acted on the succeeding layer, and this on the next, and so on, so that all the layers appear almost at the same time; but the proximal ones are somewhat nearer the definitive form than the more distal ones.

Pflüger also points out that certain exposed surfaces can organize new material only in one direction. In other words, the organized surfaces of the body show a polarization, since one side of a surface shows peculiarities not present in the other. The direction of the polarization of the regenerating surface is the cause of the direction of the new growth. For this reason a portion of an animal cannot produce the entire animal.

Pflüger has not taken into account some of the most conspicuous facts of regeneration — facts known even at the time at which he wrote. His explanation fails completely to account for those cases where a piece of an animal, hydra, for example, changes its form into a new small hydra. Moreover, Bonnet had shown that when Lumbriculus is cut in two, only a few new segments are added at the anterior end of the posterior piece. The same thing is true for the earthworm, in which, if the piece be cut off behind the fifteenth segment, no new reproductive region is ever developed. Pflüger's view leaves the absence of this region unaccounted for.

Weismann, in his book on the *Germ Plasm* published in 1893, elaborated an hypothesis of regeneration. Weismann's central idea is not different from Bonnet's (1768). Both believe in preformed germs. The differences in their views result largely from the application of modern cell doctrines to Weismann's hypothesis. Regeneration is supposed to be brought about by latent cells containing preformed germs which exist in the chromosomes of the nucleus in the form of determinants. There are supposed to be cells of this sort in the leg of a newt, for instance, at every level and in all the parts. At each level the latent cells are slightly different, and each contains germs of such a sort that all the distal part and only the distal part is represented. This germ stuff, coming into action after a series of qualitative nuclear divisions, influences the part in which it is found. Further, since the new limb will itself regenerate if cut off, the further assumption is made that during regeneration new subsidiary germs are laid down at each level in the new limb. This is supposed to take place by a division of each germ into like parts (a quantitative division) after it has reached its proper position in the new leg.

This host of invisible germs, moving at the command of Weismann's imagination, is supposed to carry out the process of regeneration. No one can fail to see that the difficulty is only shifted into a region where fancy can have free play and a scientific, experimental test cannot be applied. At one blow the difficulties are overcome, and the array of mystical germs is summoned to explain how regeneration takes place.

Since regeneration occurs in some animals and not in others, and better in those forms, Weismann thinks, that are liable to injury, additional hypotheses are added. Weismann combats the idea that regeneration is the result of the inherited effect of injuries to the part. The Lamarckian conception cannot apply in this case, since it is not the use or disuse of a part that is in question, but its power to regenerate after injury. In this I agree with Weismann, for it is no more evident how a series of injuries in succeeding generations could finally bring about complete regeneration than that the result should follow after one injury. If after the first injury, then, there is no need of any theory of inheritance.

Weismann believes, however, that the power of regeneration is under the guidance of "Natural Selection." His argument, as far as I understand it, is this: Of all the animals injured in each generation, those that regenerate better are more likely to survive; and since in each species certain organs are more liable to injury than others, the selection will take place mainly in respect to these parts; hence they possess the power of regeneration. Other organs of the body not subject to frequent injury do not show the power of regeneration, either because, having once had it, it has been gradually lost (through panmixia), or because the process has never been acquired in these organs. It may be pointed out, in passing, that since the limb of a newt and the tail of a tadpole regenerate at every level, and regenerate the kind of limb peculiar to that particular species, we must conclude, on Weismann's hypothesis, that this power has been acquired through selection for every possible level. The demand made on our credulity is enormous.

Weismann has made other statements in his book on the *Germ Plasm*, and in a later paper ('99) entitled "Regeneration: Facts and Interpretations," that are worth examining. In the former he says: "It may, I believe, be deduced with certainty from those phenomena of regeneration with which we are acquainted that *the capacity for regeneration is not a primary quality of the organism, but that it is a phenomenon of adaptation*." Again: "Hence there is no such thing as a general power of regeneration; in each kind of animal this power is graduated, according to the need of regeneration, in the part under consideration." "We are, therefore, led to infer that the general capacity of all parts for regeneration may have been ACQUIRED BY SELECTION [1] *in the lower and simpler forms*, and that it has gradually decreased in the course of phylogeny,

---

[1] The italics are my own.

in correspondence with the increase in complexity of organization ; but that it may, on the other hand, be increased by *special
selective processes* in each stage of its degeneration in the case
of certain parts which are physiologically important and at the
same time frequently exposed to loss."

There are many points in these quotations that are, I think,
open to criticism, but it would lead me too far were I to attempt
to discuss them. Yet I cannot let this opportunity go by without calling attention to another point raised by Weismann in
his later article. He says: " It may not have occurred to
Morgan that the changes in the structure of a species may
have kept pace with the changes in the conditions of its life ;
yet this is a presupposition of the hypotheses of natural selection, and is, indeed, its *conditio sine qua non.* Hermit crabs
have certainly possessed the power of regeneration 'from the
beginning'; but may they not have inherited it from their ancestors, the long-tailed forms, which possess it to this day, and have
need of it for all their appendages, since all are liable to injury ?
And cannot, nay must not, these in their turn have inherited it
from their ancestors, the sessile-eyed crustaceans, and so on,
through the whole crustacean pedigree, back to the unknown
annelid-like ancestors of the class ? . . . It seems almost as if
Morgan ascribed to me the view that the capacity for regeneration must be built up anew for each species—must be inscribed,
so to speak, on a *tabula rasa;* my view, however, is that here,
as in all transformations, nature started with what was already
present, and by modifying it brought about adaptation to new
conditions. The assumed general power of regeneration in the
lowly ancestors of the crustaceans would thus gradually have
adapted itself to the changes in the body and to the new conditions resulting from these changes as well as from other
causes ; it would have become localized and specialized. . . .
As in the course of time the appendages of the different
body-segments became more widely differentiated in adaptation
to different functions — giving rise to antennæ, jaws, walking-
legs, or swimmerets — the predisposition to regenerate in certain
parts of the body slowly varied also; and thus, not indeed at
the same rate, but not lagging very far behind, the adaptation

of the capacity for regeneration followed the adaptation of a limb to a new function."

Weismann had, in his book on the *Germ Plasm*, committed himself to the statement that the germ plasm for regeneration is different from that which brings about the development of the egg. He believed that it was necessary to make this hypothesis to account for the appearance of the so-called "ancestral organs" that are sometimes (?) regenerated. Combining this statement with what has been quoted above, it will be clear that as a new species evolves from an older one its regenerative germ plasm must also change if the animal regenerate a part like the one lost. Hence each species must acquire the power of regenerating its own particular kind of structures. This opinion I did ascribe to Weismann, and still suppose he holds to it. I did not imply that regeneration "must be inscribed, so to speak, on a *tabula rasa*." The capacity to regenerate the parts of the old species would be present in the new one, but the new species must acquire through natural selection of favorable variations the power to regenerate the new structures that have arisen through egg variations. I have quoted at length this argument of Weismann to show where we are landed by the results of his speculation. He argues for a double process of natural selection for each species that can regenerate, and is led into this position by the assumption that seems quite necessary on the preformation hypothesis that regenerative germs exist in the egg independent of the germs of embryonic development.

I find in this whole argument only an attempt to shift the difficulties of the problem back to the unknown ancestors of present forms, just as the difficulties of other parts of the problem are also shifted back upon the unknown germs that exist preformed in the egg or in the parts that can regenerate. Weismann does not, perhaps, realize the difference between himself and those whom he somewhat scornfully calls "the younger investigators." The problems that they are trying to solve are those that Weismann also tries to answer, but "the younger investigators" base their interpretations on the assumption that when a change takes place a sufficient cause for the

change is to be sought in the organ itself and in the external
conditions surrounding that organ. They are not content to
rest their "explanations" on "the phyletic origins" of the
changes. It is not necessary to deny the theory of descent,
but it is unsafe and in many cases unscientific to base "causal
explanations" on an imaginary line of ancestors. It is certainly unprofitable to shift our difficulties back to these historic
forms, and most unfortunate to find our "explanations" also
resting on the same shadowy past.

In a masterly essay, entitled "Stoff und Form der Pflanzen-
organe," Julius Sachs has considered the question of regeneration in plants, and has outlined an hypothesis to account for
the phenomena. Sachs bases his view on the conception that
the form of a plant is the outcome of its chemical structure;
that whenever the form changes there has been an antecedent
change in the material (Stoff). Sachs vigorously combats Vöch-
ting's idea that there exists in the organism a polarization of
every part, and that this polarization is a directive agent that
determines the kind of regeneration that takes place. No less
earnestly does Sachs protest against the metaphysical conception of many morphologists, expressed or implied, *viz.*, that for
each species of organism there is a form that tends to express
itself and controls the development of each part. According
to Sachs there are no such formative forces in the organism,
but all changes are brought about by differences in the chemical composition of the "Stoff," and this leads to the development of the form peculiar to that material. For example, the
flower-buds of plants are produced, not because of some innate,
mystical force that causes the plant to complete its typical form,
but because some substance is made in the leaves that, flowing
into the undifferentiated growing point, there acts on the material of the growing point and changes it into that sort of stuff
from which a flower develops.

Applying this idea to regeneration, Sachs supposes that in
the plant two substances are being produced; one of these is
a leaf-forming substance, the other produces roots. When a
piece of the stem is cut from a plant these two substances contained in it flow in their respective directions, and bring about

the production of new leaves and new roots. Sachs made numerous experiments which showed, he thought, that the direction of the flow of these substances is determined by the action of gravity; the leaf-forming substances flowing upwards, and the root-forming downwards; the direction of the flow being thus determined by some factor outside of the plant itself. It had been shown that if twigs of the willow be planted with the distal end in the ground, new roots arise from the end in the ground, and leaves from the free end. This result follows, Sachs thinks, from the direction of flow of the root-forming and leaf-forming substances.

In considering Sachs's view, it will be well, I think, to keep apart the two ideas, that specific substances produced in the plant bring about the change, and that these substances may be transported from one part to another in definite directions. We might think of the transportation taking place in a given direction as due to some peculiarity of the substance itself, or of the tissues of the organism, or, as Sachs supposed in this case, as the result of some outside influence.

It is interesting to notice that this idea of a transportation of specific stuffs goes back to Bonnet. The latter imagined head-nourishing and tail-nourishing stuffs to flow, respectively, forwards and backwards, and, acting on the germs in those parts, determine the kind that develop. Sachs does not introduce the idea of preformed germs, and correspondingly simplifies his hypothesis.

The idea of specific substances determining the regeneration of a part is, in my opinion, one deserving of very careful consideration. We have seen that the idea first suggested itself to Bonnet when searching for an explanation of the development of a new head or tail from the same region of Lumbriculus. A head developed if the exposed part lay at the anterior end of a piece, and a tail if the exposed part lay at the posterior end. Bonnet said that something must awaken the one or the other kind of germ, since both were assumed to be present at every level. Hence the idea of two specific stuffs. It is evident, of course, that the difficulty is only shifted from the germs to the stuffs, for no such stuffs were known, much less their migration in definite directions. Nevertheless, the assumption gave a formal explanation of the phenomena.

I shall next consider, in the light of the hypothesis of transportation of specific stuffs, the results of certain experiments that bear on the question.

When an oblique piece is cut from a planarian, as indicated in Fig. 1, *A*, I find that the new head develops not from the center of the piece but far up on one side, Fig. 1, *B*; and always on that side that lay nearer to the head of the planarian. The new tail develops out of that part of the posterior side that was originally nearer to the tail of the planarian, Fig.

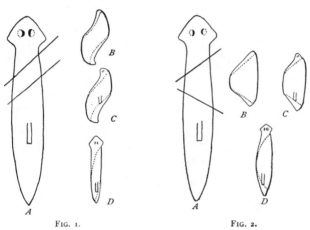

FIG. 1. FIG. 2.

1, *B, C.* The new pharynx appears in the middle of the piece, with its long axis at first in the direction of the old long axis; but as the old part changes its form to produce the body of the new worm, the pharynx comes to lie symmetrically with respect to the new head and tail, Fig. 1, *D*.

If a piece is cut from the planarian by making two oblique cuts that form an angle with each other, as shown in Fig. 2, *A*, a new head develops, as before, at the most anterior part of the anterior cut surface, and a tail at the most posterior part, Fig. 1, *B*. In this case, both head and tail lie on the same side of the piece. The new pharynx appears in the middle, but not in line at first with either the new head or tail.

These experiments show that the position of the new head

and of the new tail is determined by the obliquity of the exposed part. The nearer the cut approaches a plane at right angles to the long axis, the nearer the new head and tail will be in the middle line.

If we assume that specific substances bring about the development of the new head and tail in these pieces, and also assume that the one substance flows towards the most anterior end of the piece, and the other towards the most posterior end, we should have an apparent explanation of the results.

A further application of this hypothesis may be made to certain phenomena in the regeneration of the earthworm. If the worm (*Allolobophora fœtida*) is cut posterior to the middle into two pieces, the posterior piece produces in many cases, at its anterior end, a new tail and not a head. If we assume that in the tail region only tail-forming substances are present, and little or no head-forming substance, we might offer this supposition as an explanation of the result. But a difficulty arises in connection with the idea of transportation of this stuff, for on the hypothesis it should flow backwards instead of forwards. However, the unusually long time before this sort of regeneration begins might be utilized to save the transportation theory. We might assume that the tail-forming material flowed posteriorly, but after a time so much will have accumulated that it may extend forward even to the anterior end, and there start the regeneration. It is obvious that this is a forced interpretation and that the result is not in accordance with the transportation idea.

This example of heteromorphosis in the earthworm is due, it appears, to influences within the piece itself. This seems true also for those cases, described by Herbst and myself, in which an antenna is regenerated in place of an eye in some of the crustaceans. This case also does not harmonize well with the transportation hypothesis, although, as I shall try to show later, it might be explained on the assumption of specific materials in the parts themselves.

There are other cases of heteromorphosis in which, as shown by Loeb, the influence that determines the kind of regeneration comes from the outside. In certain hydroids regener-

ation is influenced by the orientation of the pieces with respect
to gravity; in others to light ; in others to contact.   In those
cases in which gravity is the determining factor, we could
readily imagine that the transportation of head-forming or tail-
forming substances is brought about by the action of gravity.
In the other cases in which light or contact has an influence
on the regenerating part, it is not easy to see how a stimulus
of this sort could bring about the transportation of material,
although we can readily imagine that either factor acting
locally might cause the production of some substance which, in
turn, might act on the new tissues.

These illustrations would seem, in several cases at least, to
harmonize well with the stuff-hypothesis, less well, perhaps,
with the idea of its transportation.   On the other hand, one
does not have to look very far to find other cases to which
the stuff-transportation hypothesis does not apply, or applies
badly.

In the first place, in many of the lower animals regeneration
does not take place by the development of new tissues, but by
a remoulding — morpholaxis — of the entire piece into a new
form.   This side of the question has been almost entirely neglected by those who have proposed hypotheses of regeneration,
and yet it seems to me that just here we find some of the most
important phenomena.   A protozöon cut into pieces makes as
many new individuals of small size as there are nucleated
pieces.   The size of the new individual is, within certain limits,
in proportion to the size of the fragment, and it develops —
regenerates — not by forming new material at the cut surfaces,
but by remoulding the entire piece into the characteristic
form.   If a piece is cut from a hydra, it bends together and
makes a sphere or cylinder out of which a new hydra is formed.
There is no evidence of the formation of new tissue, unless the
tentacles are formed in that way ; but in other hydroids, *viz.*,
Tubularia and Parypha, even the tentacles are known to
develop out of the old cells.   In planarians a small amount of
new tissue forms at the cut ends, and out of this a new head
and a new tail develop, but the old piece changes over to form
the body of the new individual.   It is in the higher forms alone

— echinoderms, molluscs, annelids, and vertebrates — that we find regeneration taking place only by the addition of new tissue at the cut ends.

In those forms that regenerate by morpholaxis we must find a theory that can account for the changing over of the old piece into a new whole. To assume two formative substances might account for the changes at the two ends, but not for the rest of the piece.

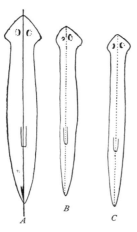

FIG. 3.

There are still other difficulties for the stuff-hypothesis. If a planarian be cut in two longitudinally, each half forms a new individual. Along the entire length of the cut edge new tissue appears. The new pharynx appears along the border between the old and the new tissue (Fig. 3, *A*, *B*). It lies at first quite unsymmetrically in the new worm, but the new side continues to grow and the old side to get narrower. The result is that although the new worm is only as broad as the half from which it developed the pharynx lies at last in the middle plane of the body (Fig. 3, *C*).

Shall we assume that side-forming substances are present, which, being transported laterally, bring about the development of the new side? If so, we should have to assume that at each point the substance must be different from that elsewhere, for a different structure is formed. I believe that few persons would like to assume the responsibility for such a view.

One further objection. If the foot of a newt is cut off, only the foot develops; if the cut is made through the forearm, then the forearm and foot are both regenerated; if the limb is cut off through the humerus, then all distal to that point is renewed. If we assume a leg-forming substance, the assumption is insufficient to account for the differences in the result at each level. If we assume a different kind of substance flowing from the body for each level, then the hypothesis becomes extremely complicated, and we might as well fall back at once on the Bonnet-Weismann theory of preformation.

There is another experiment on planarians that has a direct bearing on Bonnet's hypothesis. If a planarian be almost entirely split in two, leaving the halves connected only at the anterior end (Fig. 4), two new heads may develop at the most anterior end of the cut edges (Fig. 4). Van Duyne, who first carried out an experiment of this sort, found two heads developing, and he interpreted their development as due to a process of heteromorphosis. I have repeated this experiment a number of times with the same results, but I think there is a simpler and more obvious way to account for the development of the new heads. They appear at the sides of each half, as they would do were a long piece cut from the side of the body; but in the latter case the result is not due to heteromorphosis. In the former case the two new heads are, after their formation, prevented from being carried forward by the presence of the old head. This interpretation is in harmony with the results of several other

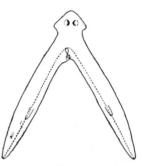

FIG. 4.

experiments. The bearing of this experiment on our present examination is obvious. Two new heads develop, although the old head is present. If the development of the new heads is due to the presence of head-forming substances, as Bonnet supposed, how could they develop as long as the old head is present to use up these substances? The objection might not apply with as much force if the transportation theory did not include the using up of head-forming substances in the old head.

Other examples might be cited, but those given above will suffice, I think, to show the improbability of the stuff-transportation theory, or, at least, *the results show that it cannot be universal in its application* to the phenomena of regeneration. The assumption of head-forming and tail-forming stuffs is too general to explain the results. We have, however, clear evi-

dence of the fact that the chemical composition of the various
organs of the body is different, and I think it is not going too
far to claim with Sachs that in a sense their form is the outcome of their composition.   We have also many observations
to show that during regeneration like parts form like.   In the
earthworm, for instance, the new ectoderm comes from the old,
the new digestive tract in large part from the old one ; the new
nerve cord comes, in part at least, from the old one and in part
from the ectoderm.   In other words, the specific character of
the old cells may be handed over to the new ones.   If we
choose to think of a formative substance in each kind of cell,
we could make this the basis of our interpretation of the results ;
but since we do not know of such formative substances, it is
safer to rest our claim simply on differences in the chemical
substance of the cell itself.   In other words, we only complicate
our view by assuming formative substances acting on an indifferent medium—the protoplasm—and determining its changes.
We need not deny that this might sometimes happen, and, in
fact, several cases that Sachs has cited seem to be due to some
such action ; but, in general, it is not necessary to make this
distinction, at least not in most cases where regeneration takes
place.[1]

The assumption, in those cases in which regeneration takes

---

[1] By composition of a cell or of any part of the body, we may mean either of
two things.   The fundamental substance may be thought of as everywhere the
same, *i.e.*, indifferent, and its action is directed by grosser substances (formative
substances), or we may mean by composition the entire substance without making
any distinction between the substances of which it is made up.   For the latter alternative we may use the terms " chemical " or " molecular " structure or " constitution,"
and retain the term " formative substance " for the former alternative.   For example,
we may think of a growing point as made up of indifferent protoplasm, and its fate
determined by the kinds of formative substance carried to the growing point (Sachs).
When new substance is added during regeneration at a cut surface, we may think
of the new cells as at first indifferent, and their fate determined by the formative
substances transported to or acting at the cut surface, or we may think of the new
cells as having a molecular structure like that of the cells (or parts) from which
they have come.   This molecular structure might be thought of as changing later,
at least so far that at different levels in the new part it changes in its relation to
what is around it, and forms a new whole.   The change would be due to some
other factor or factors, such, for instance, as the form of the new part or its
relation to the old part.

place by the development of new tissue, that the new cells have inherited their specific nature from the cells from which they have arisen, meets with no serious objection. But can we explain all the results as the outcome of this inheritance? I think not. For instance, the assumption will not explain why the new part assumes a form that is often different from that part of the body from which its cells have been derived. For example, if the first five segments be cut from the anterior end of the earthworm, five will come back; but if more than five are cut off, still only five come back. Now the new cells will in the latter instances be derived from parts of the worm that are quite different in their structure from the fifth or sixth segment, and yet from any level of the anterior end of the body only the five segments develop. The assumption of formative substances, or of a molecular structure in the new cells, might account for the fact that the new ectoderm comes from the old, the new muscles from the old ones, endoderm from endoderm, etc.; but that would leave the main problem unexplained, *viz.*, the development of a new head. This example shows also that Pflüger's hypothesis is insufficient to account for the result; for, according to his hypothesis, we should expect that at every level the whole of the missing part should be replaced, and not merely five segments, at whatever level the cut is made.

We see that other factors are also at work. Let us see if we can account for some of these. The digestive tract, at first ending blindly in front, pushes out into the new part until it comes in contact with the ectoderm. Its further growth forward is prevented, not simply by meeting a mechanical obstacle, but by what we may call, for want of a better term, a stimulus received from the point of contact. At this point the ectoderm now turns in to form, as Hescheler has shown, the buccal chamber. We might assume that the endoderm acted in turn as a stimulus on the ectoderm, and the invagination then followed; but it is probable, from certain results of Rievel and of Hazen, that the invagination would take place, even if the endoderm did not come in contact with the ectoderm. The new nervous system extends forward from the old one. In coming to the anterior end, its further growth forward is prevented by the ectodermal covering, and the cord then divides and turns upwards around the digestive tract to form the commissures and brain. Still the most essential point is unaccounted for, *viz.*, that the cord divides at the anterior end, for were the new part a tail and not a head, the division would not take place. These examples will suffice to show that the phenomena are too complicated to be accounted for by the form of the new part ; however, the form of the new part may be one of the factors, but not the only one. The obvious fact that cannot be left out of account is that the new part is a head long before the cells undergo their definitive differentiation ; and it seems to be owing to the new substance being from the start a head, that many of the results take place. Nothing is easier at this stage of the analysis than to drop into metaphysical ideas and speak of vital factors or formative forces, etc. I have not escaped this pitfall altogether, for I pointed out on one occasion that the process appeared as though guided by intelligence. " I mean that what we call correlation of the parts seems here to belong rather to the category of phenomena that we call intelligent than to physical or chemical processes as known in the physical sciences. The action seems, however, to be intelligent only so far as concerns the internal relations of the parts, etc." [1] The reactions to stimuli that are amongst the most common phenomena of living things is what I had in mind, and it is true that at present we cannot explain them as the result of known chemical or physical properties of matter, but I do not think that therefore I was justified in calling them intelligent processes, even in the broadest use of the word, for we thereby fall into the error of attempting to explain simpler processes by more complicated and less well-understood ones.

We are confronted, then, with the question : What is it that gives to the new part the structural character or chemical composition to form a new head? If we say its form, *i.e.*, its closed dome-shaped figure, we meet with the objection that this same form is almost universally present where the new part is present as a knob. However, since in each such knob of new cells the

[1] " Some Problems of Regeneration," *Biol. Lect.*, 1898.

cells have a different molecular structure from those derived from other regions, we might, by combining the two assumptions, escape from the dilemma. On the first assumption, that the cells of each new layer or new organ have received their molecular character from the old part, and in each kind of tissue are all more or less alike, we are driven to assume, on the second assumption, that the form of the new part determines subsequent changes in the molecular composition, and the material is so changed at each point that its arrangement produces the foundation of a new head.[1]

Before we follow further this line of thought let us examine those cases in which the entire piece is changed over into a new organism. If a piece is cut from the middle of the body of hydra, it closes in at both ends and a cylindrical form is the result. If the piece owed originally its characteristic form (*i.e.*, as a piece of the body of hydra) to its molecular structure, it would have this same character after separation ; and in giving rise to a new hydra it does not develop new tissue at the two ends like that lost, but, on the contrary, after a short time the old piece itself assumes the form of a new hydra. If we hold to Sachs's view, that wherever a new form arises there has been an antecedent change in the material, we must conclude that in the piece of hydra the material has rearranged itself into a new whole. The cells do not materially change their position, but each goes over into a different part of the body from that of which it formerly made a part. Since, however, as in the case of Lumbriculus, each cell may develop into a part of either the anterior or posterior end of the new hydra, according to the level at which the piece was cut off, we must account for the one or the other change. What factors can we suppose bring about this result ? In the first place, however the piece is

[1] The specific substances that the new cells have brought with them from the old parts may in some cases determine the kind of new organ rather than the relation of the new organ to the rest of the body (through, of course, its area of contact with the old body). In this way a new eye may develop in the hermit crab when the distal end of the old eye stalk is cut off, and not an eye but an appendage when the stalk is cut off at its base. The development of a tail at the anterior end of a posterior piece of the earthworm would be due to the specific character of the new cells dominating the development.

cut out, the material at one end will have been more anterior in position in the original piece than the material at the other end. This difference is sufficient, theoretically, to account for the different results at the two ends. I do not assume that a polarization is present, because we do not know enough about any such principle, if it really exists at all, to make it of any service. My assumption rests simply on the basis that at each level the material is somewhat different from that at every other level. These considerations give us a sufficient basis to build up an idea of how the development of the piece may be thought of as taking place. As soon as the piece has been removed and its ends have closed in, it is possible to think of a rearrangement of the molecular structure taking place throughout the whole mass. Since the two ends are different, we can imagine their differences to become greater and greater in the same directions in which they differed at first. If the expression is pardonable, the anterior end becomes more anterior, and the posterior more posterior; and this influence extending to the intermediate regions, they too change in their respective directions. As a result, the material of the new piece assumes the molecular arrangement characteristic of a hydra. Then the differentiation begins that changes the entire piece into the new individual.

Let us not hesitate to push this view to its logical conclusions and ask in what part of the material does this change take place. Does each cell change and through simple contact with its neighbors bring about a change in them, and so from cell to cell? In other words, is the result an intercellular reaction? While it might be possible to look at the result as brought about in this way, still, I think there are several important reasons why we must regard the change as more fundamental than that involving only the cells as units. I cannot take up at this time my reasons for so thinking, but I may point out that since in the Protozoa and in the eggs and embryos of other forms changes similar to those I have described take place, we seem to be forced to the conclusion that the change is in the whole protoplasm and is probably a molecular change of the protoplasm.

To repeat; when a piece is cut out of the body of hydra a molecular change takes place in the protoplasm of such a sort that the entire mass is changed over into a structure that represents in its structural basis a new hydra. It is this molecular change that, dominating the subsequent development, seems to control it, and gives us the impression of formative processes at work. On my view, the formative processes are only the expression of the physical, molecular structure that has been assumed by the piece.

It will be seen at once how the same conception may be applied to those cases of regeneration in which a knob of new tissue appears and the missing part, or a portion of it, is regenerated. After the new cells have been formed they will have the same relations to each other and to the old part that the cells have in the isolated piece of hydra. The new part does not form a new whole, but only a part, simply because it is connected with the old part by the same kind of molecular union as are the cells of the new part to each other. Whether all the missing parts, as in the limb of the newt, or whether only a part, as in the head of the earthworm, will develop out of the new material, will be determined by the volume of the new material that forms, and its relation to the structural peculiarity of the new part. This sounds vague, perhaps mystical, but I think it can be given a real meaning. If, for instance, five segments are cut off from the anterior end of the earthworm, the new material that forms suffices to make five segments, but if ten be cut off, the new material is still only sufficient, owing to some molecular peculiarity, to make five.[1] In the limb of a newt we must suppose that at every level enough new material is formed at first to make possible the formation of any part of the limb.

This analysis of one of the problems of regeneration has been undertaken in order to see if it is possible at the present time to construct an hypothesis that can bring under one point of view many isolated observations. It is offered as a working

[1] There is a difficulty, perhaps, in accounting for only two segments coming back when only two are cut off, and not five, but if it can be shown that less new material is formed the difficulty is avoided.

hypothesis or as a possible point of view, and not as an elaborated theory of regeneration. In fact, I think it would be a mistake at the present time to attempt to construct a final theory, for, if I have been in the least successful in this discussion of the problem, I hope to have made it clear that the process of regeneration involves many factors. It is obvious that until we have analyzed the problem into its component parts it would be ridiculous to attempt to formulate a theory of regeneration. It is not difficult to show that there are in reality many factors in the process quite different in kind. For instance, the closing in of the exposed surface seems to be due to some sort of cytotropism in the cells as well as to other factors; the production of new cells from the old ones must be due to another set of processes, as well as their migration out over the exposed end; the form of the piece and its size are also factors; the specific character of the cells derived from the old ones is still another problem; and finally, if my analysis is sound, the subsequent molecular arrangement taking place throughout the new part is one of the most important changes that take place. I have laid emphasis only on this latter characteristic because it seems to me that it is just this change that comes nearer to what we mean by regeneration than any of the others that I have named; they all enter into the problem, however, and none of them can be neglected, but to find a theory to account for them all at once would be, I think, extravagant to attempt and probably disastrous in its results.

I O

JACQUES LOEB
1859-1924

Brought up in Germany, Loeb received his M.D. in 1884 from the University of Strasbourg, then moved to Würzburg. He spent 1889–1890 at the Naples Station and there encountered *Entwickelungsmechanik*. He also became attracted by the possibilities for a career in the United States because of his friendship with American researchers there. After marrying an American, in 1891 he moved to the United States and taught at Bryn Mawr, then at the University of Chicago. As Whitman had done with Wheeler, he made major efforts to keep Loeb at Chicago, but lack of support from the administration and Loeb's disagreements with members of his department of physiology there ultimately induced Loeb to leave for the University of California. He spent summers at the MBL, where he served as instructor, head of the physiology program, and Trustee.

Loeb's work on animal tropisms began with study of caterpillar's reactions to light. Tropisms rather than special instincts direct movement, Loeb maintained. Despite skepticism, Loeb gained considerable support for his views during the 1890s, largely buoyed by his extensive experimental work. Increasing criticism by Thorndike, Jennings, and others around 1900 did not change Loeb's views. He continued to develop his physicochemical, mechanistic program of biology, maintaining further that sufficiently full understanding could lead to control of life. A second line of research on what stimulated cell division and the associated development took Loeb to analysis of artificial parthenogenesis. Changing saltwater concentrations could stimulate cell division and development, he discovered, at a time when such an idea was "in the air" and a number of researchers were pursuing similar lines of exploration. As a result, Loeb's successes received considerable and immediate attention. Heredity and evolution seemed relatively unimportant for development while immediate proximate physicochemical conditions seemed primary, for Loeb. His viewpoint influenced some, including Thomas Hunt Morgan, but did not ultimately carry the day with the majority of researchers at the MBL.

JACQUES LOEB.

ON THE NATURE OF THE PROCESS OF FERTILIZATION

JACQUES LOEB

I.

LEEUWENHOOK demonstrated in 1677 the existence of sper matozoa. It was about one hundred and sixty years before biologists convinced themselves that these spermatozoa were no parasites. In 1835 K. E. von Baer was still of the opinion that the spermatozoa had nothing to do with the process of fertilization. The parasitic conception of spermatozoa was finally done away with by Wagner's demonstration that only those animals are capable of fertilizing eggs whose sperm contains spermatozoa. Very soon afterwards histologists showed the origin of spermatozoa from cells.

Leeuwenhook was of the opinion that the spermatozoön represents the future embryo. On the other hand, it was not difficult to notice that the embryo in fishes, amphibians, and birds develops from an egg furnished by the female. The question arose as to what was the homologous element in the female of mammals. In 1672 De Graaf discovered the follicle in the ovary of mammals, and in 1827 von Baer discovered the mammalian egg cell.

The next problem that was solved in this branch of science was the relation of the sperm to the egg. Many scientists, among them De Graaf, had assumed that no direct contact between egg and sperm was necessary, that something volatile in the sperm, the aura seminalis, was sufficient for the act of fertilization. Contrary to this, Jacobi showed (1764) that fish eggs can only be fertilized by bringing the sperm into direct contact with the eggs, and Spallanzani showed the same for the frog. He succeeded in producing artificial fertilization of mammals by introducing sperm into the vagina. But even Spallanzani did not realize that the spermatozoa were the essential element in the sperm. In 1824 Prévost and Dumas proved this by filtering the sperm, and demonstrated that the sperm whose spermatozoa had been retained by the filter lost its power of impregnating the egg. These observations established the fact that the spermatozoön has to come in contact with the egg in order to bring about fertilization.

The next step was the observation made by Barry in 1843 that the spermatozoön actually enters into the egg. This observation was confirmed ten years later by a number of authors, Meissner, Newport, Bischof, etc. It is rather remarkable that it was one hundred and sixty years after the discovery of the spermatozoön and the follicle before the fact was recognized that the spermatozoön has to enter the egg in order to bring about fertilization. Had the biologists during these one hundred and sixty years lost their interest in the investigation of this problem? This was certainly not the case, but they spent their energy not in fruitful research, but in speculations and controversies which were admired by their contemporaries and made their authors famous, but which were a mere waste of time. History has since repeated itself in other fields of biology. The outcome of the facts gathered concerning the process of fertilization was four apparently different theories of fertilization, which, however, have much in common.

The first theory of fertilization is a morphological one. According to this theory, it is the morphological structure of the spermatozoön which is responsible for the process of fertilization.

The second theory is a chemical one. According to this theory it is not a definite morphological or structural element of the spermatozoön, but a chemical constituent, that causes the development of the egg. Against this second view Miescher has raised the objection that his investigations showed the same compounds in the egg and the spermatozoa. I do not

think that this objection is valid.   We know that simple variations in the configuration of a molecule have an enormous
effect upon life phenomena.   This is shown among others by
the work of Emil Fischer on the relation between the molecular configuration of sugars and their fermentability.   When
Miescher made his experiments he was not familiar with such
possibilities.   Moreover, Miescher was not able to state whether
the spermatozoa contain enzymes or not.

A third theory was a physical theory (Bischof).   This theory
assumes that a peculiar condition of motion exists in the spermatozoön which is transmitted to the egg and causes its development.   It should be said, however, that this idea is not so
very different from the chemical conception, because it assumes
exactly the same for the spermatozoön that Liebig assumes for
the enzymes.   Liebig thought that the enzymes owed their
power of producing fermentation to the motions of certain
atoms or groups of atoms.

The fourth conception is the stimulus conception, which was
originated by His.   According to this conception the egg is
considered as a definite machine which if once wound up will
do its work in a certain direction.   The spermatozoön is the
stimulus which causes the egg to undergo its development.   It
is to be said in connection with this stimulus conception that
the main point at issue is omitted, as to whether the stimulus
carried by the spermatozoön is of a physical or a chemical character, and in this way, of course, the stimulus conception is
nothing but a disguised repetition of the chemical or physical
theory of fertilization.

All these theories are so vague that we do not need to be
surprised that none of them has led to any further discovery.
If we want to make new discoveries in biology, we must start
from definite facts and observations, and not from vague speculations.   Among these observations the most important are
those on parthenogenesis.   It had been observed for a long
time that the unfertilized egg of the silkworm can develop parthenogenetically.   It was, moreover, known that plant lice can
give rise to new generations without fertilization.   The most
impressive fact concerning the parthenogenesis of animals was

contributed by Dzierzon, who discovered that the unfertilized eggs of bees develop and give rise to males, while the fertilized eggs give rise to females. Similar conditions seem to exist in wasps. It is, moreover, certain that a few crustaceans show parthenogenesis.

A beginning of parthenogenetic development had been observed in the case of a great many marine animals which develop outside of the female in sea water. It was found that such eggs when left long enough in sea water may develop into two or three cells, but no further. (2) On the other hand, in ovaries of mammals now and then eggs were found that were segmented into a small number of cells. These facts and the occurrence of a certain class of tumors in the ovary, the so-called teratomata, suggest the possibility of at least partial parthenogenesis in the eggs of mammals. But all these phenomena were considered to be of a pathological character. It must be, however, admitted that we cannot utilize these facts with any degree of certainty for the theory of fertilization, as in this case certainty can only be obtained by the experiment. It was not until very recently that such experiments were made.

II.

Eight years ago I observed that if the fertilized eggs of the sea urchin were put into sea water whose concentration was raised by the addition of some neutral salt they were not able to segment, but that the same eggs, when put back after they had been in such sea water for about two hours, broke up into a large number of cells at once instead of dividing successively into two, four, eight, sixteen cells, etc. Of course it is necessary for this experiment that the right increase in the concentration of the sea water be selected. (3) The explanation of this fact is as follows : The concentrated sea water brings about a change in the condition of the nucleus which permits a division and a scattering of the chromosomes in the egg. As soon as the egg is put back into normal sea water it breaks up into as many cleavage cells at once as nuclei or distinct chromatin masses had been preformed in the egg. Morgan tried the same experiment on the unfertilized eggs of the sea urchin, and found that the unfertilized egg, if treated for several hours with concentrated sea water, was able to show the beginning of a segmentation when put back into normal sea water. A small number of eggs divided into two or four cells, and, in a few cases, went as far as about sixty cells, but no larva ever developed from these eggs. (4) Morgan had used the same concentration of sea water as Norman (5) and I had used in our previous experiments. I had added about 2 grams of sodium-chloride to 100 c.c. of sea water. Norman used instead of this 3½ grams of $MgCl_2$ to 100 c.c. of sea water, and Morgan used the same concentration. Mead made an observation somewhat similar to Morgan's upon Chætopterus. He found that by adding a very small amount of KCl to sea water he could force the unfertilized eggs of Chætopterus to throw out their polar bodies. The substitution of a little NaCl for KCl did not have the same effect. (6) While continuing my studies on the effects of salts upon life phenomena, I was led to the fact that the peculiar actions of the protoplasm are influenced to a great extent by the ions contained in the solutions which surround the cells. As is well known, if we have a salt in solution, say sodium-chloride, we have not only NaCl molecules in solution, but a certain number of NaCl molecules are split up into Na ions (Na atoms charged with a certain quantity of positive electricity) and Cl ions (Cl atoms charged with the same amount of negative electricity). When an egg is in sea water, the various ions enter it in proportions determined by their osmotic pressure and the permeability of the protoplasm. It is probable that some of these ions are able to combine with the proteids of the protoplasm. At any rate, the physical qualities of the proteids of the protoplasm (their state of matter and power of binding water) are determined by the relative proportions of the various ions present in the protoplasm or in combination with the proteids. (7) By changing the relative proportions of these ions we change the physiological properties of the protoplasm, and thus are able to impart properties to a tissue which it does not possess ordinarily. I have found, for instance, that by changing the amount of sodium and calcium ions contained in the muscles of the skeleton we can make them contract rhythmically like the heart. It is only necessary to increase the number of sodium ions in the muscle or to reduce the number of calcium ions or both simultaneously.[1] On the basis of this and similar observations I thought that by changing the constitution of the sea water it might be possible to cause the eggs not only to show a beginning of development but to develop into living larvæ, which were in every way similar to those produced by the fertilized egg.

There seemed to be three ways in which this might be accomplished. The first way was a simple change in the constitution of the sea water without increasing its osmotic pressure. The second way was to increase the osmotic pressure of the sea water by adding a certain amount of a certain salt. The third way was by combining both of these methods. The first way did not lead to the result I desired. All the various artificial solutions I prepared had only the one effect of causing the unfertilized egg to divide into a few cells, but I was not able to produce a blastula. I next tried the effects of an increase in the sea water by adding a certain amount of magnesium-chloride. In this case I had no better results than Morgan. Very few eggs began to divide, but these did not develop beyond the first stages of segmentation. I then tried the combination of both methods. The osmotic pressure of ordinary sea water is roughly estimated to be the same as that of a $\frac{5}{8}$ nNaCl solution or a $\frac{10}{8}$ nMgCl$_2$ solution. I found, after a number of experiments, that by putting the unfertilized eggs of the sea urchin into a solution of 60 c.c. of $\frac{20}{8}$ nMgCl$_2$ solution and 40 c.c. of sea water for two hours the eggs began to develop when put back into normal sea water. Such eggs reached the blastula stage. I do not think that anybody has ever seen before such blastulæ as resulted from these unfertilized eggs. As these eggs had no membrane, the amœboid motions of the cleavage cells led very frequently to a disconnection of the various parts of one and the same egg, and the outlines of the egg became extremely irregular. The blastulæ

[1] It is due to the Ca ions of our blood that the muscles of our skeleton do not beat rhythmically like our heart.

showed, as a rule, the same outline as the egg had in the morula stage. It was, moreover, a rare thing that the whole mass of the egg developed into one blastula. The disconnection of the various cleavage cells led, as a rule, to the formation of more than one embryo from one egg. The results were in a certain way similar to those I had obtained when I caused the fertilized eggs of sea urchins to burst. In such cases a part of the protoplasm flowed out from the egg but was able to develop. These extraovates had no membrane, and of course showed some irregularity in their outlines, but the irregularity in this case was far less than that observed in the unfertilized eggs of my recent experiments. But although I had thus far satisfied my desire to see the unfertilized eggs of the sea urchin reach the blastula stage, I was not able to keep these eggs alive long enough to see them grow into the pluteus stage. They developed more slowly than the normal eggs, and died, as a rule, on the second day.

It was my next task to find a solution which would allow the eggs to reach the pluteus stage. I found that this can be done by reducing the amount of magnesium-chloride and increasing the amount of sea water. By putting the unfertilized eggs for about two hours into a mixture of equal parts of $\frac{2 0}{8}$ nMgCl$_2$ and sea water, the eggs, after they were put back into normal sea water not only reached the blastula stage, but went into the gastrula and pluteus stages. The blastulæ that originated from these eggs looked much healthier and more normal than those of the former solution with more MgCl$_2$. Of course as these unfertilized eggs had no membrane it happened but rarely that the whole mass of an egg developed into one single embryo. Quadruplets, triplets, and twins were much more frequently produced than a single embryo. The outlines of each blastula were much more spherical than in the previous experiment. These eggs reached the pluteus stage on the second day (considerably later than the fertilized eggs do). Thus I had succeeded in raising the unfertilized eggs of sea urchins to the same stage to which the fertilized eggs can be raised in the aquarium. I have not yet succeeded in raising the fertilized eggs in my laboratory dishes beyond the pluteus stage.

Though I do not wish to go into the technicalities of these experiments, I must mention a few of the precautions that I took in order to guard against the possible presence of spermatozoa in the sea water. The reader who is interested in this technical side of the experiments will find all the necessary data in my publication in the *American Journal of Physiology* (8). Here I only wish to mention the following points : —

1. These experiments were made after the spawning season was practically over.   2. Bacteriological precautions were taken against the possibility of contamination of the hands, dishes, or instruments with spermatozoa.   3. The spermatozoa contained in the sea water lose, according to the investigation of Gemmill (9), their fertilizing power inside of five hours if distributed in large quantities of sea water.

4. We have a criterion by which we can tell whether the egg is fertilized or not in the production of a membrane. The fertilized egg forms a membrane and the unfertilized egg has no distinct membrane. None of the unfertilized eggs that developed artificially had a membrane.

5. With each experiment a number of control experiments were made. Part of the unfertilized eggs were put into the same normal sea water that was used for the eggs that did develop. None of these eggs that remained in normal sea water formed a membrane or showed any development, except that a few of them were divided into two cells after about twenty-four hours.   6. I made another set of control experiments by putting a lot of eggs of the same female into a solution which differed less from the normal sea water than the one which caused the formation of blastulæ or plutei from the unfertilized eggs. In this case it was shown, that although these eggs received the same sea water as the ones which developed, and although they were injured less than the ones which developed, yet not one single egg formed a membrane or reached the blastula stage. If the sea water had contained any spermatozoa these eggs should have reached the blastula stage.[1]

[1] Through other control experiments I convinced myself that a treatment of eggs or spermatozoa with equal parts of a $\frac{2}{8}$ nMgCl$_2$ solution and sea water diminishes the impregnability of the eggs and annihilates the fertilizing power of spermatozoa in a very short time.

Hence, as in nine different series of experiments these results were confirmed, we may assume that by treating the eggs for two hours with a solution of equal parts of a $\frac{2}{8}$ nMgCl$_2$ solution and sea water we can cause them to develop parthenogenetically into plutei.

III.

What conclusions may we draw from these results? If we wish to avoid wild and sterile speculations, I think we should confine ourselves to the following question: What alterations can be produced in an egg by treating the same for two hours with a solution of equal parts of $\frac{2}{8}$ nMgCl$_2$ and of sea water? Even in this regard we can only give a very indefinite answer, which, however, will have to be in the following direction: The bulk of our protoplasm consists of colloidal substances. This material easily changes its state of matter and its power of binding water. It seems probable that changes of these two qualities are mainly responsible for muscular contraction and perhaps amœboid motions. Among the agencies that cause changes of these physical qualities we know of three that are especially powerful. The one is specific enzymes (trypsine, plasmase, etc.). The second is ions in definite concentration. The concentration varies for various ions. The third agency is temperature. In our experiments it is obvious that only the second possibility can have been active. I do not consider it advisable to enter into theoretical discussions beyond these statements. The next question that should be raised would be whether the spermatozoa act in the same way. It is true that the spermatozoön contains a considerable proportion of salts, especially K_3PO_4, but it may contain enzymes or it may contain substances which have similar effects upon the physical qualities of the colloids, like the three agencies mentioned above.

In the last volume of these lectures I pointed out that it is impossible to derive all the various elements that constitute heredity from one and the same condition of the egg. (10) Our recent experiments suggest the possibility that different constituents of the egg are responsible for the process of fertilization and for the transmission of the hereditary qualities of the male.   While we are able to produce the process of fertilization by a treatment of the unfertilized egg with certain salts in certain concentrations, we cannot hope to bring about the transmission of the hereditary qualities of the male by any such treatment.   Hence, the inference must be that the transmission of the hereditary qualities of the male and the agency that causes the process of fertilization are not necessarily one and the same thing.   I consider the chief value of the experiments on artificial parthenogenesis to be the fact that they transfer the problem of fertilization from the realm of morphology into the realm of physical chemistry.

BIBLIOGRAPHY.

1. The historical data are taken from Hensen's Physiologie der Zeugung. (Hermann's Handbuch der Physiologie.   Bd. vi.)

2. HERTWIG, O.   Die Zelle und die Gewebe.   p. 239.   Jena, 1893.

3. LOEB, J.   Experiments on Cleavage.   *Journ. of Morph.*   Vol. vii. 1892.

4. MORGAN, T. H.   The Action of Salt Solutions, etc.   *Arch. f. Entwickelungsmechanik.*   Bd. viii.   1899.

5. NORMAN, W. W.   Segmentation of the Nucleus without Segmentation of the Protoplasm.   *Arch. f. Entwickelungsmechanik.*   Bd. iii. 1896.

6. MEAD, A.   The Rate of Cell-Division and the Function of the Centrosome.   *Woods Holl Biol. Lect.*   1898.

7. LOEB, J.   On Ion-Proteid Compounds and their Rôle in the Mechanics of Life-Phenomena.   *Amer. Journ. of Phys.*   Vol. iii.   1900.

8. LOEB, J.   On the Artificial Production of Normal Larvæ from the Unfertilized Egg of the Sea Urchin.   *Amer. Journ. of Phys.* Vol. iii.

9. GEMMILL.   The Vitality of the Ova and Spermatozoa of Certain Animals.   *Journ. of Anat. and Phys.*   1900.

10. LOEB, J.   The Heredity of the Marking in Fish Embryos.   *Woods Holl Biol. Lect.*   Boston, 1899.

EPILOGUE
Paul R. Gross

Professor Maienschein gives proper emphasis, in her eloquent introduction to this volume, to that goal of parsimony which so animated the founding luminaries, the first generation of biologists who did research and taught at the MBL: the making of One Biology. Biology was to be unified under a small number of principles, equally applicable to morphology and physiology, to heredity, embryology, evolution, to distal as well as to proximal causes. She notes that the effort did not really succeed. By the end of the great decade of the nineties, Whitman had left embryology, become a behaviorist, and was withdrawing from his exhausting service as scientific leader, manager, and political conscience of the MBL. Its leading and sometimes (politely) warring intellectuals were going their separate ways. And, as Maienschein observes, new issues were surfacing in biology, some of which were not addressed there.

I doubt, however, that the first generation's leaders were really as sanguine as they represented themselves to be. Moreover, their original program did to a remarkable degree succeed, although it required the better part of a century for the system of generalizations that would unify heredity, evolution, development—even behavior—to emerge.

The founders were surely aware that for the prior two hundred years the learned literature had been growing rapidly (at a compound rate, in fact, of 6 to 7 percent). Perhaps they did not dwell upon that fact, nor upon the likelihood that such growth would continue. But continue, of course, it did. According to the number of scientific publications, which does, it turns out, measure the number of active investigators, the body of knowledge doubles roughly every decade. Thus the literature today is several hundred times as large as it was when these lectures were being delivered, written down, and argued over. Most members of the biological community of the 1890s knew one another or, at least, what others were doing. Today it is common for the biologists of a single university to be subdivided as groups that do not, and cannot, intercommunicate.

And yet a set of principles capable of unifying the radically alternate points of view of the nineties—rigid predeterminism versus isotropism or radical epigenesis, hereditarianism versus proximate physicalism—such a set has in fact emerged. It is in the context of this stunning growth of biology, and in the context of the unique focus upon fundamentals that characterized the MBL community and that shaped the next generation of questions (such as, what are genes and how do they work?), that we must judge the achievements of the MBL, and the value of these splendid lectures. The test is simple: the realization of Whitman's dream of a unified science is the extent to which current lectures, even the Friday Evening Lectures of the MBL, lineal descendants of those included in this volume, deal with "morphology" and "physiology," with heredity and development, with cytology and physics and behavior, under a limited set of generalizations to which all can agree.

The point needs no laboring. Does such a set of generalizations exist? Of course it does. Among the few that comprise the set are four with which even a beginning biologist would be thoroughly familiar: the invariance of biochemistry, underlying a diversity of form and function over the whole *scala naturae*, with the lemma that the chemistry of "protoplasm" is *ordinary* chemistry; the encoding of heredity in particulate elements, which are themselves molecules or specific aggregates of molecules—of DNA; the omnipresence of history, engraved upon the hereditary material and hence upon its products, that is, of evolution as the reason why biological processes are of one kind, rather than another; and, finally, the fact that the hereditary material, in the nucleus of eukaryotes and as unsequestered DNA of prokaryotes, is affected continuously by extrahereditary information: from the cytoplasm, whose contents reflect its own recent history, and from the exterior world.

As to the role played by the MBL in the emergence of these generalizations, that is a question for historians to answer, I hope, in detail over the next few years. The answers should contain lessons for the next generation of organizers. Organization proves to have been very important, in the 1890s, for the MBL and for American biology. It will be even more important for the nineties of this century, since the enterprise is orders of magnitude larger, more complex, and more expensive.

I cannot trace properly, in these paragraphs, the origins of these powerful unifying ideas of modern biology from the biological lectures of the 1890s. Nor can I do justice to the labyrinth of historical fact, competitions for power and influence, that is the story of the MBL's role in their creation and acceptance. This much, however, can be said, even in an epilogue: The special power of these finely crafted, sophisticated lectures, and of the institution and community they so accurately represent, is in their very direct address of fundamentals. This was Whitman's doing, at least at the beginning. He was anxious that the evening lectures be accessible to educated nonspecialists. He believed that although the biological work must itself be painstaking, detailed, and to the highest degree technically expert, the important ideas to be derived from it could and should be communicated to thinkers who are not necessarily expert in the same way.

It was his influence that caused these lectures to respond so regularly to the same few, urgent issues. The lectures, and the many occasional papers and informal discussions of which we have record, were philosophical in the best sense. For that reason, the set of questions and conflicts of principle that followed emerged in an organized way. Every student at Woods Hole knew, by the early years of this century, what the issues were. Leaders of the first generation were agreed upon them, if not upon their resolutions. Thus when they went their separate ways, it was in pursuit of one or another of the fundamental questions that had exercised them all earlier.

Loeb, radical epigeneticist and physicalist, set out to prove that the methods and rules of ordinary chemistry do apply to the most characteristic "colloidal" entities of protoplasm, the proteins. From his efforts and those of his students, first at the MBL and then at many other places, there came the beginnings of protein chemistry and enzymology. Wilson and Morgan, more and more convinced that the role of "heredity" in development needed seriously to be defined, led two branches, curiously intertwined, of the attack on the mechanism of heredity after the rediscovery of Mendel: cytogenetics and formal genetics. The real home of the chromosome theory of heredity—the localization of Mendel's genes to the chromosomes—was the MBL. The effort to bring behavior, that most complex manifestation of the structure living things inherit and recreate in each generation, under biological rules, began at the MBL with Whitman, and was advanced from a very different point of view by Loeb. Specific interest in nervous systems was established early for that reason, and it grew year by year, so that the MBL was ripe for the emergence of a true neurobiology, not because of the Woods Hole squid, but because the communal mind recognized the long but unbroken pathway from a zygote to an act.

In the years surrounding World War II, leadership in biochemistry had returned definitively to Europe, with triumphant discoveries in Germany, England, and Scandinavia, in the field of intermediary metabolism. When distinguished biologist-refugees from Nazism came to America, they found homes, positions, and friendship across this nation; but the MBL was their spiritual home. It was so because the intellectual ground had been prepared. Hans Neuberg, Otto Meyerhof, Otto Loewi, Leonor Michaelis, and Albert Szent-Györgyi taught and worked at the MBL, and their presence stimulated a new generation of biochemists, who were to become leaders, despite the gigantic expansion of the field, which made it impossible for any one laboratory or institution to host even a significant fraction of the influential thinkers.

"These lecturers were preoccupied," a superficial reader of this volume might say, "with a few problems, rather parochial and technical, for embryologists. Their arguments couldn't have made much difference." Not so; they certainly did. The continuing argument about heredity—versus environmental signals acting upon a homogeneous protoplasm as the underlying mechanism of becoming, of biological development—set the stage for the large effort that was to bear fruit only after the 1950s. That fruit was borne is evident to anyone who attends lectures in biological development, normal or pathologic, here or anywhere else, today. One biology reigns. The genes are the ultimate blueprint. They act through mechanisms that are complex but ordinary chemical mechanisms. The blueprint is the product of natural selection. And there is continuous interplay and exchange of signals, not only between the genes and their surrounding cytoplasms, but between each cell and the medium that bathes it or the other cells that touch it.

And behavior? There is today a successful course of lectures and laboratory work at the MBL entitled "Neural Systems and Behavior." Its faculty profess topics that range from the physics of electrical current flow across the membranes of neural cells to the evolution of birdsong. In that program called neurobiology, whether it be the course by that name or the score of senior investigators who visit the MBL solely to do research, a common preoccupation is to identify, isolate, purify, clone, and manipulate genes that encode the proteins that make up the membrane channels through which ions—hence current—flow, and upon which various nonionic signals play, so as to effect, ultimately, behavior.

I will concede the point, though it is unnecessarily repeated, in my view, by some historians and politically minded scientists, that what scientists choose to study is influenced by large-scale social and economic pressures, so that the work and discoveries of scientists mirror the society's characteristic goals. This is not the full story, however. Particular group interactions among scientists, work styles and intellectual styles, also play a large part. Questions thought to be worth asking are usually the products of agreement by some sort of elite.

The early MBL was such an elite, modeled at first—intellectually, and not in terms of organization or mission—upon Dohrn's *Stazione Zoologica*. But it was an elite strictly in the philosophical, not social, sense. The questions deemed important by this elite gave rise to all the great questions of twentieth-century biology. There might well have been different questions. Biology could easily have taken a basically different turn. But because men and women such as the authors of these lectures met every summer, were in communication the rest of the year, and because the United States became a scientific power in the world within thirty years of their first assembly, they set the stage, once their program had taken shape, for much of what would happen later in the life sciences. The embryo of twentieth-century biology was, to apply words Whitman once used in quite another context, "if not predelineated, then predetermined." How and why that came to be is splendidly illustrated in the clarity, comprehensiveness, and, for the most part, honesty of these lectures.